Turning Off the Lights

The threat to community electricity in Sri Lanka

Steve Thomas, Jayantha Gunasekara,
Ruana Rajepakse

Published by ITDG Publishing
Schumacher Centre for Technology and Development, Bourton Hall,
Bourton-on-Dunsmore, Rugby, Warwickshire CV23 9QZ, UK
www.itdgpublishing.org.uk

© ITDG Publishing 2005

First published in 2005

ISBN 1-85339-594-3

All rights reserved. No part of this publication may be reprinted or reproduced or utilized in any form or by any electronic, mechanical, or other means, now known or hereafter invented, including photocopying and recording, or in any information storage or retrieval system, without the written permission of the publishers.

A catalogue record for this book is available from the British Library.

ITDG Publishing is the publishing arm of Intermediate Technology Development Group Ltd. Our mission is to build the skills and capacity of people in developing countries through the dissemination of information in all forms, enabling them to improve the quality of their lives and that of future generations.

Typeset by FiSH Books, London
Printed in Great Britain by Antony Rowe Ltd., Chippenham, Wiltshire

Front cover: ITDG/Steve Fisher

Contents

Acronyms	iv
Executive summary	v
Introduction	viii
Problems with the conventional electricity sector in Sri Lanka	1
Resources and consumption	1
Organization	1
The planning process and government policy	2
Problems with the electricity sector	2
Rural electrification and non-traditional energy resources in Sri Lanka	5
The Sri Lanka electrification programme	5
Non-traditional energy resources in Sri Lanka	6
Reforms to the Sri Lankan electricity industry	11
The proposals	12
The British Model and its suitability for developing countries	14
Critique of the proposals	18
Stimulating investment in Sri Lanka's electricity industry	21
The World Trade Organization and the GATS negotiations	25
The World Trade Organization	25
Progress on offers and requests	33
General arguments on the GATS	34
The Cancún Summit and subsequent developments	37
Conclusions	38
The GATS	38
The Sri Lankan electricity system	39
The GATS and the Sri Lanka electricity system	40
Appendix 1: Retreat of multinational electricity companies	41
US companies	41
European companies	45
Appendix 2: Information on the WTO and GATS	47
Appendix 3: Perceptions of the Sri Lankan electricity industry	48
The planning process	48
Consumer perceptions of the problems	48
The Reforms	49
Rural electrification	51
Appendix 4: The CEB generation plan: 2002–16	52
Appendix 5: Examples of operating micro-hydro projects	56
Kithulritiella village micro-hydro project, Perupalla, Maliboda, Deraniyagala	56
Thanthrikanda village hydro project, Thanthrikanda, Miyanawita, Deraniyagala	56
Veediyawatta village hydro project, Deraniyagala	56
Appendix 6: The Cancún negotiations	59
Notes and references	61

Acronyms

ADB	Asian Development Bank
AES	Applied Energy Services
BEWAG	Berliner Kraft-und Licht Aktiengeselschaft (Germany)
BNDES	Brazilian Development Bank
BOO	Build Own Operate
CEA	Central Environment Authority (Sri Lanka)
CEB	Ceylon Electricity Board (Sri Lanka)
CEMAR	Companhia Energética do Maranhäo (Brazil)
CEMIG	Companhia Energética de Minas Gerais (Brazil)
CEPA	Consolidated Electric Power Asia (China)
CPC	Ceylon Petroleum Company (Sri Lanka)
EDF	Electricité de France
ESC	Energy Supply Committee (Sri Lanka)
ESD	Energy Services Delivery
EU	European Union
GATS	General Agreement on Trade in Services
GATT	General Agreement on Trade and Tariffs
IFIs	International Financial Institutions
IMF	International Monetary Fund
IPPs	Independent Power Producers
ITDG	Intermediate Technology Development Group
LECO	Lanka Electricity Company
NEA	National Environmental Act (Sri Lanka)
NTPC	National Thermal Power Corporation (India)
PP&L	Pennsylvania Power & Light
PRSP	Poverty Reduction Strategy Paper
PSEG	Public Service Enterprise Group (New Jersey, USA)
PUC	Public Utilities Commission (Sri Lanka)
PV	Solar Photovoltaic
RERED	Renewable Energy for Rural Economic Development Project
RWE	Rheinisch Westfälische Electrizitätswerk (Germany)
SB	'Single Buyer' model
Swalec	South Wales Electricity Plc (Wales)
SWEB	South West Electricity Board (UK)
TXU	Texas Utilities (USA)
UKHP	Upper Kotmale Hydro Power plant (Sri Lanka)
USAID	United States Agency for International Development
VEBA	Vereinigte Elektrizitäts-und Bergwerks-Aktiengeselschaft (Germany)
VIAG	Vereinigte Industrie-UnternehmungenAktiengeselschaft (Germany)
WDM	World Development Movement
WTO	World Trade Organization

Executive Summary

The Sri Lankan electricity system is in crisis. It is unreliable and electricity prices are increasing, even though the price of electricity in Sri Lanka is one of the highest in Asia. These problems are the result of the failure of the energy sector planning system, mismanagement by the state electric utility, the Ceylon Electricity Board (CEB), and inadequate oversight of the sector by the government.

Current International Monetary Fund (IMF) and World Bank market-driven reform policies will not solve these problems, nor will they allow the Sri Lankan government to achieve its target of extending electricity supplies for the rural population from less than 50% coverage now to 75% by 2007.

After 20 years of civil war and with an economy teetering on the edge of crisis, Sri Lanka had no option but to seek financial support from the International Monetary Fund, the World Bank, the Asian Development Bank (ADB) and rich world countries. On 19 March, 2001[1] the IMF approved a loan of US$253 million and two years later on April 20, 2003 another of about US$567 million.[2]

The loans were dependent on a package of measures that included the privatization of state-owned enterprises. Central to the success of this strategy was the wholesale break-up and privatization of the state-owned electricity company, the Ceylon Electricity Board.

The package of reforms bore the hallmarks of a model of economic structural reform known as the 'Washington Consensus'. This 'one size fits all' reform package includes: privatization of state enterprises, fiscal discipline, redirection of public expenditure toward education, health and infrastructure investment, tax reform, competitive exchange rates, trade liberalization, deregulation and legal security for property rights. It is a cornerstone policy that is shared by the IMF, World Bank and other major international financial institutions.

The effectiveness of the Washington Consensus reforms is highly contested and it has been roundly condemned by many IMF and World Bank critics as a blunt instrument unsuitable for developing country economies. The World Bank President James Wolfensohn said in Shanghai in 2004 that 'the Washington Consensus has been dead for years'. "Chinese lesson in how to put food in the mouths of millions" Jonathan Watts, The *Guardian*, May 27.[3] For a relatively powerful country such as China this may be the case. China has the negotiating clout to insist on developing its own economic reform. However, a small country such as Sri Lanka is not in such a negotiating position.

In the case of privatization of public utilities even the World Bank has begun to question the way privatization has been enthusiastically forced on poor country economies. The Bank's Chief Economist and Senior Vice President Francois Bourguigon said in June 2004 that 'there was some "irrational exuberance" in recent years on the potential benefits of privatization' (World Bank (2004) 'Credible Regulation Vital For Infrastructure Reform To Reduce Poverty, Says World Bank' Press release, June 14, 2004.). He went on to say that private financing was still needed to expand services to poor people.

As a sovereign country the Sri Lankan government draws up its economic restructuring plan through what is known as a 'nationally owned Poverty Reduction Strategy Paper (PRSP)'. Sri Lanka's PRSP, called 'Regaining Sri Lanka', was completed in January 2003. Although the IMF and the Bank do not in theory impose these policies on countries, in practice the PRSP has to be approved by them. Once approved this opens the way to a flow of international loans and grants from major donors. For a country as desperate for support as Sri Lanka there is little room for manoeuvre in compiling a PRSP.

Within the PRSP Sri Lanka has been advised to restructure its electricity industry according to what is known as the 'British model'. This involves the creation of a wholesale electricity market, introduction of retail competition, de-integration, (Breaking up the industry into four component parts, generation, transmission, distribution and retail) and re-regulation.

The British model was designed to meet the needs of a modern industrialized country. The reforms were designed to take a mature industry with limited demand growth and strong complete networks and use private sector competitive disciplines to improve its efficiency.

Sri Lanka's electricity system, however, along with that of many other developing countries, has almost completely opposite needs and circumstances to an industrialized country. It has an immature industry with huge demand growth and weak and incomplete networks.

The changes needed to meet this growing demand are increased generating capacity and expansion of the grid. This will require huge capital investment – something the British model is not designed to attract.

The restructuring package is in danger of simply transferring publicly accountable nationally owned monopolies to weakly regulated foreign-owned near monopolies. This is something that the World Bank's own policy tries to prevent. Strong regulation of privatized utilities is essential if restructuring is to reduce poverty according to the World Bank's June 2004 report *'Reforming Infrastructure – Privatization, Regulation and Competition'*. The report also noted that 'unbundling' – the splitting up of the component parts of an electricity industry – 'makes the regulatory task more complex, which is likely to be a problem in environments with weak governance – as in most developing and transition economies'.[4]

The tragedy of all this is that Sri Lanka has been at the forefront of innovative ways of getting electricity to poor rural communities who have little prospect of being linked up to the grid. The World Bank has funded many of these small-scale off-grid generating plants, usually small hydroelectric systems.

The government target to increase rural electrification requires the increased use of these small-scale non-grid systems, yet government reform policies do little to promote these and will create further barriers to their uptake.

Though the Washington Consensus has been the main driver of these policies, there is a further danger looming on the horizon.

The World Trade Organization's General Agreement on Trade in Services (GATS) has the potential to erode social safeguards in the provision of electricity. The particular way Sri Lanka has structured its electricity companies means that GATS could soon apply to trade in electricity services. GATS will further entrench and extend the privatization of key services and will be irreversible.

Currently under GATS negotiations there is no exemption for decentralized community-managed services. For many poor communities community-managed services offer the most cost-effective way to access to key services. GATS could cut off such an option.

The report concludes that:

- The proposed electricity reforms will not solve the problems Sri Lanka is experiencing and will not meet the target of extending electricity supplies for the rural population to 75% by 2007.
- The 'British model' of electricity reform is not appropriate for Sri Lanka and other developing countries.
- The proposed electricity reforms could damage the prospect for decentralized power services for isolated communities.

If electricity services are to be brought under GATS negotiations industrialized countries should be more cautious about recommending unproven reform models and developing countries should resist such recommendations.

Introduction

The Sri Lankan electricity industry is facing a number of important issues that must be resolved if the people of Sri Lanka are to receive affordable and reliable supplies of electricity. Four of the most important issues are:

- The recent unreliability and increasing price of electricity supplies. These problems seem to have been the result of the failure of the energy sector planning system, mismanagement by the state electric utility, the Ceylon Electricity Board (CEB), and inadequate oversight of the sector by the government.
- Under loan conditions set by the International Monetary Fund and the World Bank, the government's response to these problems is to liberalize the industry and open it to foreign investors. This will entail the break-up and possible privatization of the CEB, replacing it with a fragmented industry operating under rules that in some cases are still not well defined and in others are unproven.
- The government has a target to extend electricity supplies for the rural population from less than 50% coverage now to 75% by 2007 and eventually to all the population. This target can only be met through the increased use of small-scale non-grid resources such as micro-hydro, yet government policy does little to promote this option and the proposed reforms may even create further barriers to their uptake.
- There is pressure on the Sri Lankan government by the developed countries, particularly the European Union and the USA, who wish to use the General Agreement on Trade in Services (GATS) negotiations to irreversibly commit the electricity sector to open access for international investors.

This report examines these four interlinked issues. While the analysis specifically relates to Sri Lanka, many of the arguments are more widely applicable. These include: methods of financing the rapid expansion of electricity demand in developing countries; the applicability of the privatized, liberalized model of an electricity industry (the British model) to developing countries; the case for developing countries to open their strategically important service sectors such as electricity to international investors under the GATS rules; and the use of small-scale isolated electricity systems to bring power to remote rural communities.

In Section 2, the use in Sri Lanka of small-scale resources to bring power to remote communities is examined. The resources available in Sri Lanka are assessed, the experience of the use of these resources is presented, and the factors that have contributed to the success of these projects are identified.

In Section 3, the proposed reforms to the Sri Lankan electricity industry are presented and analysed. The record of the 'British Model' and its appropriateness for developing countries is analysed and alternatives to the proposed reforms are identified.

In Section 4, we present the history of international trade negotiations since World War II and up to the formation of the World Trade Organization in 1995. The GATS negotiations are documented and the risks to developing countries of committing their service sectors to open up under GATS rules are discussed. The breakdown of trade negotiations at the Cancun Summit in September 2003 and the prospects for a resumption of the GATS initiative are analysed. In Section 5, conclusions on all these issues are presented.

Problems with the conventional electricity sector in Sri Lanka

The Sri Lankan electricity industry is suffering from severe problems. There have been serious shortages of supply since 1998. These were a result of delays in completing investments in power stations and infrastructure to meet growth in demand; irregular monsoon rainfall; and bad administration by the Ceylon Electricity Board (CEB), for example the purchase of plants that do not function properly. Emergency measures taken by the government in 2002 led to an increase in the amount of power available. These measures included, for example, the installation of diesel generators. Shortages were alleviated but at the cost of serious electricity price increases.

Resources and consumption

Sri Lanka has few conventional energy resources. Its primary energy demands are met by three main sources: biomass (53% – mainly fuelwood and crop residues), imported petroleum products (39%), and hydroelectric power (8%). In 2000, households – the largest energy user group – consumed 76% of the biomass energy and 39% of the electricity generated. Petroleum products were used mainly by the transport sector (62% in 2000), while other major users were the industrial sector and the power sector.

In 2001, the installed capacity in the electricity sector was 2000 MW, which generated 6520 GWh. Total sales were 4376 GWh, of which 41% was sold to household consumers, 20% to commercial users and the rest to industrial users. Installed capacity grew by 40% between 1996 and 2001, while consumption grew by 46%. Losses (commercial and technical) remained at about a third of generation. Total consumption (including industrial and commercial users) of electricity per person is 230 kWh per year, far lower than in Western countries (the UK consumes about 6000 kWh per person per year), while consumption per capita by residential consumers is less than 100 kWh per year compared to about 2000 kWh in the UK.

According to official figures,[5] electrification rates in the country vary from less than 20% of households in some districts in the North to almost 92% in Colombo. The average rate in rural areas stands at less than 47%, compared to the national average of less than 60%. The estate sector (primarily tea) has an electrification rate of 50%.

Organization

The main electricity company in Sri Lanka is the Ceylon Electricity Board (CEB), which owns about 85% of the generating capacity and the entire transmission network, and distributes power to about 2.4 million consumers. The CEB is fully nationally owned. The CEB has a turnover of Rs45 billion[6] and has more than 14 000 employees. The rest of the generating capacity is either privately owned or hired. As a result mainly of the government's emergency measures, the non-CEB element has grown strongly in recent years, from only 45 MW in 1996 to 296 MW in 2001.

The other major company is Lanka Electricity Company (LECO), which distributes electricity to about 350 000 consumers in the Western and

Southern coastal belt. It is also 100% publicly owned.

The electricity industry in Sri Lanka was established more than 100 years ago and was privately owned until 1927 when its management passed to the public sector with the creation of the Department of Government Electrical Undertakings (DGEU) in 1927. In 1951 the Electricity Act was passed to regulate the power sector. Generation and transmission were carried out exclusively by the DGEU while distribution and supply were mainly carried out by local authorities who were considered as licensees under the 1951 Act. In 1969 the CEB was established by a separate act of parliament and the functions of the DGEU were transferred to it. In 1983 LECO was established to take over distribution functions in the suburbs around Colombo and certain coastal areas. The CEB took over the remaining local authority distribution schemes.[7]

The planning process and government policy

The present Ministry of Power and Energy was created in 2001, combining all energy resources under one ministry. However there is no document yet setting out an overall energy policy for Sri Lanka. Therefore at present, the government's policy has to be gleaned from various different documents, all of which are only about the electricity industry and not all of which are consistent with each other. These comprise the:

- CEB Generation Plan 2002–16,
- 'Power Sector Policy Guidelines', Ministry of Power and Energy, November 2002,
- 'Rural Electrification Policy', Ministry of Power and Energy, November 2002, and
- recent acts of parliament designed to restructure and regulate the power sector.

The CEB Generation Plan (see Appendix 4) covers the conventional sector, but the use of such a centrally operated optimising methodology (the methodology uses a computer model of the system to minimize the cost of operating the electricity system)would not be consistent with the market philosophy of the proposed reforms. The Rural Electrification Policy is discussed in Section 2, while the Power Sector Policy Guidelines and the recent acts of parliament are discussed in Section 3.

The Energy Supply (Temporary Provisions) Act No. 2 of 2002, designed to alleviate the acute power crisis experienced in 2001–2 and in force until 21 March 2004, is also relevant. This is discussed in Section 1.4.3.

Problems with the electricity sector

Capacity shortage

Increases in generating capacity have failed to keep pace with the growth in demand, and successive failures of the monsoon rains have reduced the reliability of Sri Lanka's main power source – hydro. The decision-making process on new plants is unsatisfactory. Planned power projects, both thermal and hydro have been subject to long delays due to factors such as legal disputes, public opposition and government indecision. Although there is an approval procedure in place, projects continually fail to be implemented. When objections are stated, clear-cut decisions are not taken, thereby dragging out the process at the cost of the project. There are many examples, the two most recent and controversial being the Upper Kotmale Hydro Power plant (UKHP) and a coal power plant variously projected for the west and east coast and the south-east.

The resulting instability of the system, together with the expense of emergency power generation using short-term alternatives, has adversely affected domestic and industrial consumers. The shortfall has been aggravated by technical problems that have reduced equipment reliability, for example, failures with the gas turbine (supplied by Fiat Avio) at Kelanitissa have been highlighted in several newspaper reports. Steam turbines installed in the 1960s should have been retired, but are still essential to the system.

In recent years, the frequency and duration of power shortages reached untenable levels. This necessitated emergency purchases of mostly diesel-fuelled plants. The high cost of power from such plants resulted in a sharp increase in electricity prices. By 2002, the cost of producing electricity had increased by 24%, to Rs8.97 per unit,[8] and tariffs went up by about 37%. Sri Lanka's electricity tariffs are reported to be among the highest in the world.[9] Increased prices led to a reduction in consumption which, coupled with the early arrival of rain, left the government having to purchase costly but unused emergency power, contributing further to the CEB's financial crisis.

There is a widespread perception that the CEB, despite having the potential to provide a good service, has been plagued by problems of mismanagement which have not been addressed over the years and which has aggravated the power crisis.

The main elements to the mismanagement of the CEB are said to be that:

- its size and monopolistic role leave room for corruption and institutional arrogance,
- its lack of assertiveness and political interference hinder planned activities,
- customer relations are poor,
- there is a lack of strong leadership, proper direction, and accountability, and
- they do not use their technical expertise appropriately.

Impact of the shortages on industry

The recent power shedding and tariff surcharges have adversely affected the growth and competitiveness of the private industrial sector with long-term effects on productivity and therefore on national economic growth.

Prolonged periods of power cuts and tariff surcharges have:

- reduced productive time, affecting production quantity and the reliability of businesses to deliver goods on time;
- increased the cost of production as industries are paying more for their power due to the rise in tariffs for grid electricity or are having to purchase generators and fuel. All industries have felt the effects of the rising costs, but energy-intensive industries and medium and small-scale industries who are less able to absorb these increases have particularly suffered,
- reduced competitiveness, especially in export markets as competitors in other countries are able to produce the same goods at lower cost,
- discouraged investment – local and international – due to high interest rates, reduced financial resources and the high cost of production, and
- pressured the industries to cut back and reduce their workforces.

The impacts are being felt in both urban and rural areas. For rural industries such as smallholder tea factories and leaf suppliers, the effects on the population are more severe as alternative employment options are limited.

Responses to the crisis

The Energy Supply (Temporary Provisions) Act No.2 of 2002 was the main response to this crisis. It appears to be the only legal document at present that brings energy under one common authority. The Energy Supply Committee (ESC) established by this act has jurisdiction over functions relating to all types of fuel and renewable energy, and took over certain functions of the Ceylon Petroleum Company (CPC)[10] as well as the CEB. The ESC also has the task of formulating a national energy plan.

Among its other functions, the ESC was empowered to purchase power from the private sector and to fix tariffs. Under its management the prolonged regime of power cuts that had been imposed in late 2001 was rapidly ended, but there was an increase in tariffs as an unprecedented amount of emergency power was purchased.

The operation of the National Environmental Act and the provisions of the Code of Criminal Procedure and the Penal Code relating to Public Nuisance were suspended in relation to projects approved by the ESC. The long delays in implementation of power projects due to public protests and legal battles apparently influenced the thinking of the legislature in this regard. The ESC is required to obtain an environmental evaluation from the Central Environment Authority (CEA) prior to approving a project, but the recommendation of the CEA is not binding on the ESC.

The ESC is chaired by the secretary to the Treasury and comprises mainly secretaries to various ministries, the chairpersons of the CEB and CPC, and senior officials from the main investment promotion agencies. Its composition has been criticized for lacking independent professionals who can bring a technical (engineering), scientific (valuation) and natural resource management (environmental) perspective. It also lacks representation from important sectors affected by power and energy decisions, such as industrial and domestic consumers, tourism, transportation, and irrigation.

A number of shortcomings in the role of the ESC have been identified, including that:[11]

- the ESC itself is subject to directives from other sources, hence the decision-making process has not significantly improved,
- there are still planning delays,
- decisions made by ESC are mainly based on financial considerations, with insufficient scientific input,
- there is an apparent lack of a clear plan for its activities, and
- the suspension of environmental and public nuisance laws was a short-sighted move not justified under principles of democracy and good governance. For example, the ESC exercised its power in the case of a hired diesel power plant, overriding the recommendation of CEA.

Rural electrification and non-traditional energy resources in Sri Lanka

One success story for the Sri Lankan electricity industry in recent years has been the use of small-scale generation sources to bring power to communities that cannot economically be connected to the network. These systems not only bring major welfare benefits to the communities that they serve, they are also environmentally sound and represent resources that the communities can control. The use of small-scale generating technology to bring power to isolated communities is not just of relevance to Sri Lanka: 1.6 billion people worldwide do not have access to electricity, and use of such systems is a policy that many developing countries can follow and has the potential to bring major welfare benefits to poor communities.

The Sri Lanka electrification programme

According to official figures,[12] electrification rates in the country vary from less than 20% of households in some districts in the North to almost 92% in Colombo. The average rate in rural areas stands at less than 47%, compared to the national average of 60%. The estate sector has an electrification rate of 50%. There were estimated to be about 1.8 million households in rural areas and about 150 000 households in the estate areas still needing electricity at the end of 2002, with the numbers expected to increase with population growth.

Rural electrification has been a priority of successive governments and a policy document on this issue was published in November 2002. The policy states that 75% of rural areas will be electrified by 2007.[13] It is envisaged that this target will be achieved with private sector and civil society participation. However it is also envisaged that not more than 80% of all households can be finally connected to the national grid.[14] This means that off-grid sources will have to be developed for the remaining 20% (see Box 1 for the case for the need for decentralized power worldwide).

Plans laid out in this policy state that targets will be reached through:

Box 1: The need for decentralized power supply worldwide

Access to electricity is not a luxury for poor communities. Global poverty will not be reduced without energy to increase production, income and education, create jobs and reduce the daily grind involved in just surviving. Over 1.6 billion people do not have access to electricity (27% of the world's population). In South Asia only 40% of the population has electricity, and in sub-Saharan Africa the electrification rate is only 22.6%. Four out of five people without electricity live in rural areas. Private utilities will not extend networks to areas where it is unprofitable, unless governments subsidize them to do so.

There is significant evidence around the developing world to show that decentralized, community-based electrification will be essential for providing electricity to many of the world's poorest people. In China there are over 70 million people without electricity, mostly in remote, sparsely populated areas with limited access to roads, markets and services. In Peru, the government has admitted that those with no access to electricity include populations in rural, remote localities where 45% of the population live in poverty. Mozambique has the lowest rate of electrification in sub-Saharan Africa, at 7.2%, and only 1% of the rural households has grid connection.

In rural areas where grid extension would be extremely costly and populations are too dispersed for conventional distribution, it can be shown that decentralized electrification is a viable and cost-effective option for many remote communities.

- a level playing field in rural electrification,
- an enabling regulatory framework,
- cost-reflective electricity tariff setting,
- third party access to the networks, and
- a subsidy mechanism for rural electrification.

However, there is little evidence in the proposals to reform the electricity sector that these mechanisms are prominent in the minds of policymakers.

Non-traditional energy resources in Sri Lanka

While Sri Lanka has limited indigenous conventional fuel resources, it is well endowed with a range of non-conventional resources, particularly hydraulic, biomass and wind energy. The scale and location of these resources makes them particularly relevant for isolated systems. The economics of such options depend not only on the extent and nature of the resource, but also on financing arrangements. Relative to conventional fossil fuel options, such resources often have high construction costs, but low operating costs. This means that the overall cost of power from such sources is particularly dependent on the cost of capital. Where the equipment has to be imported, it will also be dependent on currency exchange rates.

The four main resources relevant to Sri Lanka are small or micro-hydro, wind, solar and dendro-power.

Small/micro-hydro

Small and micro-hydro schemes are suitable for remote areas and do not require the construction of large dams. The projects can be built with local equipment and technical expertise, thereby developing a local industry and reducing costs. The environmental impact is significantly less than that of large hydro or fossil fuel power stations. The schemes use little water, allowing for the generation of electricity even during drought periods in most areas. They can later be connected to the national network. The main disadvantage is that the schemes are resource limited and may generate only a small amount of power. As demand grows, a different resource may be needed to supplement supplies.

Micro-hydropower generation is a viable technology that has already been used to deliver significant amounts of energy to the most remote villages with hydro resources in Sri Lanka (see Appendix 5 for examples of operating projects). The viability of off-grid micro-hydro technology for rural electrification was first demonstrated in Sri Lanka by the Intermediate Technology Development Group (ITDG) in 1991. Micro-hydro units are fully managed by the village communities. A formal society is formed (Electricity Consumer Societies), to handle the main functions such as operation, maintenance, and tariff collection. It is also responsible for management, maintaining discipline and making sure all the members uphold the procedures and standards set up by the society leadership.

In order to create an enabling environment for wider uptake of the micro-hydro off-grid rural electrification option, a series of measures were carried out by ITDG, including capacity-building at a national level on project design, implementation and evaluation; capacity-building at a local manufacturer/supplier level; capacity-building at provincial council and local NGO level; and awareness creation at policy formulation level.

A recent survey carried out by the Energy Conservation Fund operating under the Ministry of Power and Energy in Sri Lanka revealed that there are 161 micro-hydro units in operation in good condition providing the basic electricity requirements of 3687 households. The total capacity is 1622 kW. There are also 24 units with a total capacity of around 385 kW that are not functioning. There are 16 units under construction with a planned capacity of 238 kW.

Provincial councils have been playing a very active role in taking this technology to the most remote parts of the country. The Uva, Sabaragamuwa and Southern Provincial Councils have endorsed the technology by including it in their policies and by setting up separate ministries for renewable energy. Every year they allocate funds for micro-hydro schemes. So far provincial councils have partly or fully invested in 70 off-grid micro-hydro schemes.

The World Bank-funded Energy Services Delivery (ESD) project developed 56 off-grid micro-hydro power plant with a total capacity of 574 kW providing electricity for 2800 households during its period of implementation (March 1997 – June 2002). The World Bank's assessment of this project stated:[15]

> 'The ESD Project has supported the installation of 350 kW of village hydro systems serving 1732 beneficiary households. This result exceeded the original ESD target of developing 250 kW in capacity, but was lower than the 2000 rural households anticipated at appraisal. Although the capacity target was exceeded, lower number of households benefited due to possible underestimate of household demand. The appraisal estimate was 100 W/household but in practice the demand is found to be about 200 W/household. As against a targeted output of implementing 20 systems, a total of 35 systems were implemented during the course of the project. A further 49 projects are at various stages of completion and have been transferred to the follow-on Renewable Energy for Rural Economic Development Project (RERED). These projects were approved given the importance of maintaining the momentum of village hydro market take off. The completed project costs show an average capital cost of US$2060/kW. This is comparable to the economic capital cost estimated at appraisal of US$2023/kW.'

ITDG carried out a survey to assess the off-grid micro-hydro potential in Sri Lanka for the ESD project in 2000. The results show an estimated meteorological potential of 41 490 kW in 1023 sites in 10 districts of the Uva, Central, Sabaragamuwa and Southern provinces. Past experience shows that this capacity is sufficient to provide the basic electricity power requirement of approximately 187 630 households on the basis of 200 W/household.

Wind

Wind power technology is well-proven worldwide and is increasingly being exploited in developed countries. As a result, costs are falling significantly and, particularly for small machines, the equipment can readily be manufactured in local facilities. It has already begun to be used as a grid and a stand-alone resource in Sri Lanka. The CEB has a 2 MW plant in Hambantota. Wind power potential is restricted to certain areas that have adequate wind regimes (over 6 m/s annually). In Sri Lanka the areas with most potential are Hambantota, Trincomalee and Puttalam. Like solar power, back-up or power storage facilities would be needed for isolated systems if a continuous supply was required.

Following the approach of its successful experience in the micro-hydro sector, in 1988 ITDG, with the support of a team of experts, designed a small 250 W capacity wind generator suitable for rural electrification. The cost of fabrication and installation is around US$550. Initial piloting of these systems was carried out in the Southern Provinces of Sri Lanka. Along with pilot testing, local manufacturers were trained to construct small wind systems. Since then 22 systems have been installed in Sri Lanka and are in operation. The experience so far shows the full potential of small wind systems as a viable technology option to cater to the rural electrification needs.

Recently CEB carried out a wind mapping exercise in Sri Lanka with financial support from USAID. The mapping exercise was carried out by the National Renewable Energy Laboratory of the United States Department of Energy. The results show very positive meteorological potential in many parts of the island, especially in the Northern, Eastern and Southern provinces.

Solar

Energy from the sun can be converted directly into electricity using solar photovoltaic (PV) technology. Like hydro power, PV technology can be used as a stand-alone or a network-connected resource. Use of PV technology has little impact on the environment and Sri Lanka is well endowed with suitable sites throughout the country. At present, however, PV panels have to be imported and are expensive, and large-scale plants do make significant demands on land. The nature of the resource means that for isolated systems, providing uninterrupted supplies of power would require a storage device (batteries) or a supplementary source.

Solar PV was introduced to Sri Lanka in the 1980s, and since then the market has gradually grown. Unlike in the cases of off-grid micro-hydro and small wind systems, solar home systems are promoted and marketed mainly by the private sector. Currently there are about 20 000 solar home systems installed in Sri Lanka. The recently completed solar resources mapping exercise in Sri Lanka revealed a very high meteorological potential for solar energy use.

Dendro-power

Dendro-power involves the burning of biomass (trees) to produce electricity. A variety of technologies can be used, from simple combustion to gasification and subsequent combustion, perhaps in a combined cycle plant. Dendro-power is well established in countries such as Sweden, Norway, China and Thailand. Unlike hydraulic and wind options, dendro-power is not strongly constrained by local resource availability. Power can be produced in large quantities but can also provide a decentralized energy option for rural development. Use of biomass does not result in a net increase in carbon dioxide levels but there can be problems of deforestation unless production is strictly managed and monitored.

Biomass is the most common

energy source in Sri Lanka. Worldwide there are three technologies tested and proven to transform the energy in biomass into electrical energy. These are not yet proven in Sri Lanka, however, especially in small-scale operations. The Sri Lankan Ministry of Science and Technology is presently testing technologies for power generation, including a 35 kW demonstration unit in Kelanitissa. The tree crops being proposed to produce wood chips for fuel are gliricidia (used as a shade tree in the tea industry), leucena and acacia. Energy Forum, a network of experts and stakeholders in the energy sector in Sri Lanka, is in the process of testing the viability of community-based off-grid dendro thermal electricity generation.

Sri Lanka has a total land area of 6.5 million hectares, out of which 3.7 million hectares are agricultural land. Nearly one-third of these lands are under shifting cultivation in the dry zone areas. Shifting cultivations are cyclic and therefore at any given time the majority of these lands are under-used. These lands can be effectively used for fuelwood farming. The labour force in the agriculture sector can be involved in fuelwood farming since this gives them an additional source of income.

Factors affecting the uptake of off-grid electricity options

There are a number of conditions apart from resource feasibility that should be fulfilled in order for the off-grid electricity generation technologies described above to be taken up in the rural areas.

- **Community collective management capacity:** The off-grid village hydro experience has shown the viability of community-based electricity generation and management of such schemes. Communities are capable and willing to take the responsibility for managing systems that provide the basic needs for their day-to-day life. The evidence shows that this is and will be the case for the other off-grid technologies that have the potential to fulfil basic electricity requirements.

- **Technology absorption ability:** The capacity to manage the operation and maintenance and repairs locally is one of the main factors affecting the technologies' sustainability in the rural context. ITDG's experience in technology transfer has shown that basic knowledge and expertise are available in rural areas to absorb and manage technologies. There are electricians and mechanics who can be trained to handle most of the repairs and maintenance of these systems. For repairs beyond the control of the village, there is access to light engineering workshops in nearby towns with resources such as three-phase electricity and heavier equipment. If the technology transfer plans take sufficient measures to build the capacities of these individuals and enterprises, the rural communities are able to manage the technologies with resources within their reach without being dependent on Colombo or major cities.

- **Capital investments:** Many of the potential off-grid electrification systems discussed above are based on renewable supplies. The main cost will be the initial capital, then there are no running costs except maintenance and repair. The micro-hydro experience and other development interventions by many organizations have proven communities' desire and ability to contribute their labour, materials and some finances to develop their village infrastructure. The World Bank-funded ESD project facilitated uptake of off-grid electrification systems through provision of credit. According to the local banks and micro-credit institutions that disbursed credit, repayment rates are very high – almost 100%. This experience shows that the rural communities are reliable customers, proving the viability of off-grid electrification systems through credit. The Sri Lanka government's rural electrification policy acknowledges that off-grid energy technologies are the best way to achieve its electrification targets. Therefore it is necessary to back this policy by ensuring that the finance for the investment in off-grid electrification technologies is available.

Reforms to the Sri Lankan electricity industry

After 20 years of civil war and with an economy teetering on the edge of collapse, Sri Lanka had no option but to seek financial support from the International Monetary Fund, the World Bank, the Asian Development Bank and rich world countries. On 20 April, 2001 the IMF approved a loan of US$253 million, then two years later on April 18, 2003, approved another of about US$567 million.

The loans were dependent on a package of measures that included the privatization of state-owned enterprises. Central to the success of this strategy was the wholesale break-up and privatization of the state-owned electricity company, the Ceylon Electricity Board (CEB).

The package of reforms bore the hallmarks of economic structural reform known as the 'Washington Consensus'. This 'one size fits all' reform package includes: privatization of state enterprises, fiscal discipline, redirection of public expenditure toward education, health and infrastructure investment, tax reform, competitive exchange rates, trade liberalization, deregulation and legal security for property rights. It is a cornerstone policy shared by the IMF, the World Bank and other major international financial institutions.

The effectiveness of the Washington Consensus is highly contested and it has been roundly condemned by many IMF and World Bank critiques as a blunt instrument unsuitable for developing country economies.

In the case of the privatization of public utilities the World Bank has begun to question the way privatization has been enthusiastically imposed on poor country economies and whether privatization really benefit poor people.

At the launch of its report on the privatization of public utilities in June 2004, *Reforming Infrastructure – Privatization, Regulation and Competition*, the Bank's Chief Economist and Senior Vice President Francois Bourguigon said 'While there was some "irrational exuberance" in recent years on the potential benefits of privatization, the fact is that utilities in developing countries need private financing to maintain and expand services to the poor'.

The overall conclusion of the report was that 'credible regulation is essential to ensure that reforms involving restructuring or privatization of infrastructure utilities such as water, power, transportation and telecommunications improve their performance and help reduce poverty'.[16]

Though the Bank sees regulation as essential it also admits that in developing countries regulation is a problem. The report makes the point that 'unbundling' – the splitting up of the component parts of an electricity industry – 'makes the regulatory task more complex, which is likely to be a problem in environments with weak governance – as in most developing and transition economies'.[17] This admission fundamentally questions the Bank's drive to privatize and liberalize public utilities in developing countries.

Privatization and liberalization can lead to large, unpredictable swings in consumer prices. These have been damaging even in strong developed economies, such as California and Ontario, which experienced rapid increases in tariffs after liberalization and had to abandon the reforms.[18] However, for a more fragile developing economy, where many small consumers and local companies have great difficulty paying their electricity bills, the burden of unstable high prices would be very damaging.

A small developing country would also have difficulties finding the resources to re-take control of the electricity sector. It is still not clear whether the liberalization model of breaking the industry up into its main activities (generation, transmission, distribution and retail) and creating a wholesale market for power will improve efficiency and stimulate sufficient investment to ensure secure electricity supplies.

Experience in California and Ontario suggests the case is still unproven. In a small system such as that of Sri Lanka, however, there are far too few power plants for there to be any hope of creating a fully competitive wholesale market.

If this is the case then it begs the question of why Sri Lanka has been advised to reform its electricity sector in such a way.

The proposals

The Sri Lankan government is trying to tackle the causes of the recent power shortages and high tariffs by reforming the electricity industry. In 2002 it passed the Electricity Reform Act, which would result in the break-up of the CEB both geographically and by activity. Generation, transmission and distribution would be separated; the generation monopoly would be opened up, and a national transmission company and five separate regional distribution companies created. In 2003, the government announced that it would not sell the distribution companies immediately. However, adjustments have been made to the law to allow foreign ownership of utilities and the proposals only have any logic if a privately owned industry is anticipated.

Power sector policy guidelines

In November 2002 the Ministry of Power and Energy issued two documents, 'Proposed Power Sector Policy Guidelines' and 'Rural Electrification Policy' (see Section 2.1). The 'Guidelines' set out the basis for the proposed reforms. The vision statement of the 'Guidelines' envisages 'a reliable supply of grid electricity to be available to at least 80% of the population at affordable prices'.

The main elements of this policy are:

- lower prices for the consumer,
- higher level of service and supply reliability, and
- greater private sector investment.

Private sector financing for power generation is envisaged for thermal, small and medium hydro, and non-traditional energy sources. Major hydro above 50 MW is to be reserved for the state sector because of the large quantities of water involved.

Commercialization and 'corporatization' are major ingredients of this policy, although the state will presumably play a major role in the 'integrated resource planning' which is envisaged in Paragraph 5.5 of the 'Guidelines' and in ensuring 'security of supplies' which forms another plank of the policy. The tariff policy emphasises 'sound commercial principles' and price stability, although there will also be a 'commercially based allocation of costs among consumers according to the burdens they impose on the system'.

A major role is to be assigned to an independent regulator in the form of the Public Utilities Commission of Sri Lanka, which is to regulate economic, technical and safety matters.

The commercial structure

The main legislation concerning the structure is the 2002 Electricity Reform Act No. 28. The effect of this Act will be to end the CEB's monopoly over transmission and the CEB/LECO monopoly over distribution of electricity.[19] The Act envisages a geographical and functional division of the industry. The generation, transmission and distribution activities will be handled by different entities, while the distribution activity will be split into a number of regional companies.

Sri Lanka has selected the 'Single Buyer' model to determine on a day-to-day basis which power plants are operated. Under this model, power generated by both state-owned and independent power suppliers will be purchased by a system operator, expected to be the national transmission company (the 'Transco'), which will in turn sell the power to the different distribution companies. It is presently envisaged that there will be five distribution companies whose areas of operation, with one exception, will fan out from the prosperous south-west towards the rural hinterland, in a bid to equalize the value of their respective markets as far as practically possible.

These companies will initially be state-owned, but steps are to be taken to make the new companies commercially attractive. Although CEB and LECO assets will be transferred to these new companies, only a portion of the CEB's massive debt burden will be transferred. The goal is to enable the new companies to operate in a 'commercially viable environment'. Debts that are not transferred to the new entities will be transferred to a shadowy 'Z' company, but it is not clear how they will be finally discharged. Concerns have been expressed as to whether the public will eventually have to bear the burden of CEB debt-shedding.[20]

Any privatization of the electricity industry, if it results in the formation of private power companies, will also open the way for foreign ownership of such companies, as the government has already amended its exchange control regulations to allow for up to 100% direct foreign investment in utility companies.

The regulatory regime

The legislation on regulation is covered in the Public Utilities Commission Sri Lanka (PUCSL) Act No.35 of 2002. Under this, the Public Utilities Commission (PUC) is to be a multi-sector regulator for the electricity and petroleum industries[21] and other utility service industries that may be brought under its purview by law from time to time (PUCSL Act Section 1).[22] The PUC will be the licensing authority for the generation, transmission and distribution of electricity. It will have price control powers as well as other regulatory, standard-setting and consumer protection functions. The model for this commission appears to have been the multi-sector regulators found in each of the states of the USA.[23]

The PUC may advise the government on all matters concerning the industry but is also subject to general policy guidelines from the Cabinet of Ministers (PUCSL Action Sections 17 and 30).[24] Its members are nominated by the minister in charge of policy development and implementation (usually the prime minister)[25] but require approval of the Constitutional Council.

As part of its statutory duties the PUC is required to look after a number of competing interests, for example the interests of consumers, the promotion of competition, efficiency in resource allocation, safety and service quality, operational efficiency and efficiency of capital investment (PUCSL Section 14.2).[26] The degree of priority it accords to each of these concerns will no doubt depend on the policy guidelines it receives and the mindset of its members. The lobbying power of different groups of stakeholders may also be significant. The PUC is to draw up a Manual of Procedure, the draft of which it has promised to put up for public discussion.[27]

Isolated systems

Despite the government's reliance on isolated systems to fulfill a major public policy, that is to increase the level of electrification in rural areas from under 50% to 75% by 2007, there is little if any acknowledgement of this in the proposed reforms.

The British model and its suitability for developing countries

The deregulated electricity market model

While the true pioneer in electricity deregulation was Chile, with reforms dating back to 1983, it was the reforms in Britain in 1990 that created most interest and which have been used, more or less literally, as the model for other reforms. At about the same time, John Williamson of the World Bank coined the phrase the 'Washington Consensus'. This was originally a prescription for Latin American economies but has been generalized as a prescription for all countries.

The World Bank summarized the ten pillars of the Washington Consensus in its *World Development Report 2000/2001*:[28]

1. fiscal discipline,
2. redirection of public expenditure toward education, health and infrastructure investment,
3. tax reform – broadening the tax base and cutting marginal tax rates,
4. interest rates that are market determined and positive (but moderate) in real terms,
5. competitive exchange rates,
6. trade liberalization – replacement of quantitative restrictions with low and uniform tariffs,
7. openness to foreign direct investment,
8. privatization of state enterprises,
9. deregulation – abolition of regulations that impede entry or restrict competition, except for those justified on safety, environmental and consumer protection grounds, and prudential oversight of financial institutions, and
10. legal security for property rights.

Clearly, these ten points are strongly in support of the WTO's objectives in general and its GATS objectives in particular.

What is the 'British model'?

There are four essential elements to the 'British model':

- **Creation of a wholesale electricity market.** Under this element, owners of power stations would have to compete on an hourly basis to sell their power.
- **Creation of retail competition.** Consumers would be able to choose from a field of retail electricity suppliers.

- **De-integration.** To ensure that all generators and retail suppliers are able to enjoy non-discriminatory access to the network, the network should be owned (or at least managed) by companies with no interest in generation or retail supply. In some cases, generation and retail supply are kept separate. If generation and retail are carried out by the same companies, the wholesale market could become irrelevant.
- **Re-regulation.** The elements of the electricity industry that could not be made competitive, such as the high-voltage transmission system (grid) that takes power from power stations to centres of demand and the low-voltage distribution system that takes power from the grid to final consumers, would be regulated by an independent regulator.

A fifth element, privatization, that is the sale of publicly owned companies to private investors, is a frequent but not essential element. In the Nordic countries, where the system was reformed in 1991–6, ownership of the large number of companies was mostly public and change of ownership was no part of the reform policy. In the USA, most electric utilities were already privately owned before the reforms and so privatization has not been part of the reform process there.

However, while privatization is not apparently a necessary part of the British model, in practice publicly owned companies tend to move into the private sector after liberalization. In a truly competitive market, public owners have little opportunity to use ownership of an electric utility to achieve policy goals. In those circumstances, the public owners are likely to feel that they can achieve more by selling the company and using the proceeds in areas where they can have an impact. So even in the Nordic countries, there is a growing trend to private ownership of electricity companies.

For developing countries, the opening of the sector to foreign investment is also an important element. In many developing countries, national utilities find it difficult to access sufficient capital to make the investments necessary to meet growing demand. The promise of liberalization was that international companies with easy access to investment funds would enter the market and solve the problem of shortage of capital.

Does the British model work?

While all four of the basic elements (creation of a wholesale market, introduction of retail competition, de-integration and re-regulation) require massive changes to the industry, it should be the first element of the reforms, wholesale competition, which provides the major pay-off. Typically, the cost of generation represents 50% or more of the final cost of electricity and it is the promise that a competitive generation system will be more efficient than a monopoly system that has driven reforms, particularly in developed countries. If generation remains a monopoly, there is little scope for retail competition because the retail element of a bill is typically only about 5%. If there is not to be wholesale competition, there is no need to guarantee non-discriminatory access to the network so de-integration would be unnecessary. Reforms have generally led to the introduction of more explicit and more independent forms of regulation, but independent regulation of monopolies could have been introduced without other reforms.

While many reforms appear to follow the principles of the British model, in practice, few if any carry through fully all its elements. In Britain, the role of the wholesale market is very limited. The government caved in to pressure to allow integration of generation and retail supply in 1998. The industry structure quickly collapsed and it is now dominated by just five large integrated generation/retail supply companies. Most power is now generated by these integrated companies for their own consumers and does not pass through the wholesale market and most of the balance is bought and sold under long-term confidential contracts. This is hardly the hour-by-hour competition in generation that was promised at the time of privatization in Britain.

Designing an efficient wholesale electricity market has proved problematic elsewhere, most conspicuously in California, and none of the existing market designs can claim to be fully proven. While the failure in California was dramatic, in other countries, such as the UK, the problems have been dealt with by weakening the competitive element while still maintaining a façade of competition.

Retail competition has proved expensive to implement and the benefits have tended to go to large consumers with the resources, incentive and market muscle to negotiate the best deal. In Britain, the introduction of retail competition has seen a large transfer of costs from large consumers to small consumers with retail suppliers reserving their cheap electricity supplies for large consumers.

Private companies have been reluctant to surrender their network assets and, for example, in nearly all countries the distribution network is still owned by an electricity retail supply company. In Germany and parts of the USA, the transmission network is still owned by the dominant generation companies. In the regulatory activity, regulators have not been as independent of government as they sometimes claim to be. Governments generally appoint (and can dismiss) regulators and often have legal powers to over-rule regulatory decisions. Regulators often have very limited resources compared to the companies they regulate.

Differences between developing and developed countries

There is an important distinction between the motivation for reforms in developed and developing countries. For developed countries, the main rationale for the reforms was to take a mature industry with limited demand growth and strong complete networks and use private sector competitive disciplines to improve its efficiency. Often there were also a number of other motives involved with little specific connection to policy objectives in electricity. For example, in Britain, the privatization programme was also seen by the government as a way of reducing state involvement in economic activities and as a way to reduce union power (by breaking up large state-owned companies). In developed countries, the need for new investment capital was relatively small and easily satisfied under the old monopoly system.

For developing countries, the priority has often been to find a structure that would give access to new sources of capital to meet rapidly growing demand and to strengthen relatively weak networks. Economic efficiency was important, but the costs of not being able to meet demand are likely to be much higher than the costs of a degree of inefficiency. The World Bank has promoted privatization (the Washington Consensus) and it stresses what it sees as the positive effect of privatization on growth. It also refers to the problem that government-owned utilities are subject to counterproductive interference by governments and that corruption in such utilities is a problem. Others suspect that the IFIs have other motives, such as producing government income to allow them to pay debts to the IFIs more easily or generating new business opportunities for the companies of the developed countries that finance the IFIs.

Applying the British model to developing countries
The different priorities of developed and developing countries causes problems in implementing these reforms. The rationale for the British model was to create competition, so it is not well designed to encourage investment. Competition increases investment risk and, added to the intrinsic risks of developing country markets (political and economic instability), this is likely to increase costs and possibly deter investment. It seems perverse if the priority is to stimulate investment in new assets to privatize existing companies because this channels foreign money into buying existing facilities rather than into new facilities. If we look in detail at the British model, it does not appear well suited to developing countries.

Generation. For many developing countries, competition in generation is not feasible. In countries which often have relatively few power plants and a shortage of generating capacity, introducing a competitive generation market would be a recipe for market abuse by dominant generators and very high prices at peak demand periods. Experience in Brazil also suggests that in systems with a large volume of hydro-electric power, a competitive market is particularly difficult to design. In wet years, income for owners of thermal power plants will be low and a run of wet years could put the thermal power plants out of business whether or not these plants are needed to ensure secure supplies in dry seasons. Creating a market inevitably means that investment in new plants is more risky than in a monopoly market. In a market, there are losers as well as winners and potential investors, even if they are prepared to take that risk, will impose a substantial risk premium on new investments in the form of a requirement for a higher real rate of return on capital. In Britain, for example, the owners of the monopoly transmission network are allowed a 6.25% real rate of return on capital, while investors in new generating plants generally apply a test of a 15% real rate of return on new investments. If the additional charges, inevitably paid for by consumers, are not paid for by the cost savings generated by competition, it is hard to see a justification for creating a wholesale market.

Retail competition. It is generally easy to create a competitive retail market for large consumers, provided there is some form of wholesale market from which competing retailers can buy their supplies. Large consumers have the resources and incentive to negotiate hard for the cheapest supplies and the costs of allowing competition (e.g., the costs of the computer systems to allow switching and the metering costs) are small relative to the total electricity bill of a large consumer. However, small consumers tend to have little interest in shopping around for power. They do not to want to experiment with a vital purchase and they lack the skills and resources to identify the best deal. As a result, in countries in which there is full retail competition, few small consumers switch and large consumers benefit from lower prices but at the expense of small consumers. In developing countries, in which household consumers often consume very little and have difficulties paying even for this small amount, it is hard to imagine a field of retail suppliers competing over consumers that offer such low returns.

De-integration. If wholesale competition and retail competition are not feasible, it is hard to see the rationale for de-integration. Allowing the main company to own the network would not cause any competitive problems. However, separating the network would lead to the fragmentation of the industry. In developing countries, the national electricity utility is often one of the largest companies in the country and may be a valuable national resource, having the scale and resources to be a 'centre of excellence' developing national skills and capabilities, and providing a model of good practice in training and employment conditions. If the company was broken up, this would not be possible and even if the company was not broken up but sold to foreign investors, the new owners would have no interest in taking on a broader role of national skills development.

Regulation. A primary requirement for a liberalized privatized system is that it should be regulated by a body with the resources and the political power to take on international investors. The first loyalty of international investors is to their owners (or shareholders), not consumers, and their duty is therefore to maximize profits in the businesses they own. For commodities where a fully competitive market is possible, this apparent conflict of interest is not necessarily a problem because competition should ensure that consumers are not exploited. However, if the 'market' is still in effect a monopoly, or if newly created markets are not fully competitive, consumers are likely to be exploited unless an effective regulator is in place. Building up regulatory capability is not a simple task. Even in Britain, where there is long experience of dealing with international companies, at best regulators are probably only one step behind the companies in dealing with new ways to inflate profits. At least in a country like Britain, a multinational company that exploits consumers is likely to receive damaging publicity worldwide, whereas such a company's activities in a small developing country are likely to arouse little interest.

Summary. As a result of these problems, where it has been most badly applied in developing countries, the British model has often turned out to be little more than a transfer of ownership from nationally owned monopolies to weakly regulated foreign-owned near monopolies.

Critique of the proposals

The Single Buyer
There are few details of how the new

model will function in practice. One aspect that does seem to have been settled is the use of the 'Single Buyer' (SB) model as the means of determining, on an hour-by-hour basis, which power stations should be operated. Two points should be made about the SB model: first, while the SB model has been much discussed, it has not been implemented anywhere and is therefore entirely unproven; and second, there is a wide range of ways in which a SB model could be implemented.

Basically, under the SB model, a single entity has the job of buying all the wholesale electricity supplies and then selling them at standard terms to the distribution/retail companies. The job of the SB would be to ensure that wholesale electricity supplies were purchased economically and efficiently, through a mixture of buying under long-term contracts and short-term instruments.

The SB model was developed by Electricité de France (EDF), the French nationally owned electric utility. It was widely seen as a way for France to comply with the European Union (EU) Electricity Directive of 1996, which attempted to liberalize EU member states' electricity industries, while still allowing EDF to plan and co-ordinate the French electricity system. Under the EU directive, countries appeared to have three options for organizing the way in which power plants were dispatched [taken on or offline on an hour-by-hour basis]. Power could be bought and sold through bilateral deals between generators and retailers on open short-term markets (for example, a spot market like the NordPool that operates in the Nordic region); via long-term bilateral contracts between generators and retailers; or through a Single Buyer. The Single Buyer was seen as a centralized option that would allow long-term strategies (such as a nuclear power programme), through long-term contracts.

In practice, the distinction between these three different models is very blurred. From 1990–2001, Britain's power was ostensibly bought and sold through a 'Power Pool', through which all generators and retailers had to operate. All generators had to place a bid representing the price they were prepared to accept to operate. The bids were sorted by price with the the cheapest bids selected until demand was fully met. The highest successful bid set the Power Pool price and this was paid to all successful bidders regardless of how much they actually bid. Retailers had to buy all their power from the Power Pool and paid the Power Pool price. Arguably, the Pool can therefore be seen as a 'Single Buyer' operating purely on price bid. In practice, bilateral contracts were allowed and these actually dominated (more than 90%) the buying and selling of power. Generators and retailers still had to go through the Pool formally, but the terms of the contracts meant the price that was paid was set by the contract and had no relationship to the Pool price.

So, at one extreme, the SB model could be a strongly market-driven mechanism with short-term bids the determining factor. New plants would be built at the risk of the developer with no guarantee of how much power would be sold and what price would be paid. At the other extreme, it could be an instrument of central planning with plants given long-term contracts and the new capacity selected by government on strategic grounds.

In the event, France chose not to adopt the Single Buyer model and nor did any other EU member state. The option has now been withdrawn from the directive. In fact no country outside the EU has yet implemented the SB model, so it remains untested.

Privatization proceeds

Although the government claims there are no plans to privatize the new companies, the proposals only make any sense if the companies are to be privatized. The power sector reforms envisage the unbundling of the CEB and the distribution of the CEB functions amongst seven companies. CEB assets will be transferred to the new companies, while liabilities will be transferred to 'Z' company, which will remain with the government. This seems to imply that the liabilities of CEB will remain a public liability that all citizens will have to pay, even people who will never have access to the grid-connected system. Given the emphasis on the level playing field and fairness of tariffs, this proposal to load historic liabilities on all the population regardless of whether they receive an electricity supply seems paradoxical. One way to prevent such an injustice would be that if and when these companies are sold, the proceeds from their sale should be used to pay off CEB debts/liabilities.

Supplies for remote communities

There are concerns about the impact of the reforms on the poorer communities in Sri Lanka, particularly as they only focus on the main grid electricity system. Over recent years in Sri Lanka, considerable progress has been made towards establishing an environment and a market for decentralized energy supply (micro-hydro, solar, wind and bio-fuels) to supply the needs of poor and remote communities. The reforms and the current policies will have negative effects on the progress of these decentralized, community-based schemes, unless mechanisms are set in place to address them.

Sri Lanka needs an updated energy policy. The 'Proposed Power Sector Policy Guidelines' and a 'Policy for Rural Electrification' (see Section 2.1) were issued at the same time but the reforms are based only on the former. The current rural electrification policy does not take into account the power sector reforms and the implications it would have. Similarly, the 'Guidelines' do not refer to rural electrification. There is a need for a comprehensive energy policy that includes rural energy issues, and how they will be addressed. Although there are policies, there are no strategies or plans for rural electrification.

The off-grid community-owned electricity generation systems do not currently have a legal status, as only CEB and LECO are mandated to generate and distribute electricity. Despite this, these systems are recognized and promoted by the national and provincial governments.

The new policy does nothing to remove this legal anomaly and in some respects makes things worse. The reforms prevent a single license holder from both generating and distributing electricity. Exemptions are mentioned in the 'Guidelines' but it is not clear what these exemptions imply. Will the rural community-owned off-grid systems be able to apply for the exemptions? Will the exempted license holders be covered by the regulator?

One distribution company can have exclusive rights to a geographical area. This too will have implications for community-owned systems. Is it possible to have an agreement which doesn't give 'full exclusive rights', so that decentralized distribution is permissible outside the areas covered by the grid?

Policymaking for electricity is devolved. Currently, the Provincial Council and local authorities play an important role in rural electrification. There are significant numbers of off-grid systems promoted by Provincial Councils, especially in Uva, Sabaragamuwa and the Southern Province. The 'Guidelines' do not refer to this link. Will a link be established between the Public Utilities Commission (PUC) and the Provincial Council to look after the interests of communities in the promotion of village-level off-grid systems?

Mini-hydro systems that are connected to the grid use water resources that belong to the communities surrounding them, and which they directly and indirectly conserve and manage. In such a situation, if no benefits (income or electricity connections) are given in return to the community, the local community would be unfairly exploited. There is a need to ensure that part of the profits generated from these projects are invested in the economic and social development of the communities surrounding these sites.

Stimulating investment in Sri Lanka's electricity industry

A key objective of the current reforms in Sri Lanka is to bring in new sources of capital so that investments can be made in new generating capacity and in strengthening the infrastructure. This raises three questions:

- Are the proposed reforms to the Sri Lankan electricity likely to stimulate investment?
- Is there currently a field of international companies looking to invest in markets such as Sri Lanka?
- What alternatives are there to privatization and foreign investment?

Are the proposed reforms likely to stimulate investment?

The current plans for the Sri Lankan electricity industry are not entirely clear, but the main elements appear to be a de-integration of the industry into a transmission company, a number of distribution companies and a number of generation companies. At present, it is stated that there is no intention to privatize, nor is there an intention to create a wholesale market or retail competition. A regulatory body has been set up. Independent power producers (IPPs) would be encouraged to enter the market.

If there is no intention to privatize the industry and create competition, it is hard to understand the rationale for these proposals. De-integration is only necessary to ensure that competing companies have fair access to the network. Splitting up the CEB into these separate units merely fragments its capabilities. However, the key element is the expectation that IPPs would invest in Sri Lanka to meet growing electricity demand, and it is this expectation that needs analysis.

The rationale for IPPs is partly one of economic efficiency and partly to overcome shortages of investment capital. If a need for new capacity is identified and the existing utility is unable to finance it, an IPP could be brought in. The IPP would supply the finance, build, and operate the plant. In return, the company running the system would give binding commitments to buy at least a minimum quantity of the output of the plant at a pre-determined price. Competitive forces would be introduced in the process by selecting the IPP by open tender and to find the company willing to offer the cheapest electricity.

On the face of it this is an attractive model, and in the 1980s the World Bank was keen to promote IPPs as a way to improve efficiency and overcome capital shortages. However, a number of disadvantages have become apparent and the World Bank now acknowledges that in many cases the use of IPPs was too extensive.[29] To understand these problems, it is necessary to examine what conditions an international investor is likely to require.

An investor who is being asked to invest the sums of money necessary to build a new power plant (of the order of up to US$1 billion) is clearly going to require strong assurances on the income it will produce. This is likely to mean a 'take or pay' contract for 10 years or more to buy a specified amount of output from the plant and a clear predetermined price structure that leaves the plant owner with minimal risk on fuel purchase costs and inflation. Of particular importance given that most IPP developers are likely to be international companies is a requirement that income be guaranteed in a stable international currency, almost invariably the US dollar.

In return for these guarantees the local company faces little risk from construction cost overruns. If the plant does not perform reliably the owner may be required to pay compensation, although if the unavailability of a plant leads to a power cut it is unlikely that the true cost of such a power cut can be recovered – costs could be huge and could bankrupt the plant owner.

It is clear from these conditions that IPPs should only be a small element of the generation mix. They rely on the presence of a strong existing company that can provide the flexible plant necessary to meet fluctuations in demand. In developing country markets there are particular risks from over-reliance on IPPs. Developing country currencies are less stable than the US dollar and any serious decline in the value of a currency relative to the dollar will automatically be passed on to consumers as much higher power costs. Also, demand in developing countries is less predictable than in developed countries, and while it often increases sharply it could equally fall sharply in a recession. A combination of these factors – falling local currency value and falling electricity demand – led to severe problems in Pacific Rim countries such as Thailand and Indonesia after the currency crisis of 1997/98. Nationally owned electricity utilities were effectively bankrupted by the need to buy power at dramatically higher prices.[30] More recently, declines in the value of the Brazilian and Argentinean currencies (which fell by about 75% in real terms in about a year) and heavy falls in demand also led to serious problems in those countries.

If there really is no way to finance new plant construction by the existing company and as long as IPPs are only a small proportion of generating capacity, commissioning new IPPs may be a 'least bad' solution but it does bring serious risks.

Are international companies likely to invest in Sri Lanka?

When national electricity industries began to open up to international investment through IPPs in the 1980s and through privatization from the mid-1990s onwards, there seemed to be a large field of companies, particularly from the USA, but also from Europe, keen to expand outside their home markets and with ample resources to invest. From about 2000 onwards, however, there has been a progressive retreat of these companies, particularly the American ones. Now, there are no US companies looking to expand outside North America and most are selling (or abandoning) their non-US assets, often in a way that is very disruptive to the country in which they invested.

The European companies are less conspicuous in their exit, but none are expanding and there are signs of a retrenchment back into European markets. Appendix 1 details the retreat of the international electric utilities from developing country markets.

It remains to be seen whether there will be new companies stepping in to fill the vacuum left by the withdrawal or lack of interest from the American and European companies. Many of the largest utilities outside Europe and the USA are publicly owned, for example, Eskom (South Africa), CFE (Mexico), Eletrobras (Brazil), EGAT (Thailand), PLN (Indonesia) and KEPCO (Korea). Some of these are subject to break-up and reform and none seems likely to have the freedom to expand outside their home territory or contiguous markets.

In 2003, the nationally owned National Thermal Power Corporation (NTPC) of India (where it owns 21 GW of plant) began to look outside India for business opportunities in adjacent countries and in the Middle East. In October 2003, it signed an initial agreement with the Sri Lankan government to build a 300 MW plant in Sri Lanka, probably burning coal and perhaps expanding later to 900 MW. If approved by the Sri Lankan government, the project would probably be a joint Sri Lanka/Indian one. It would work on a 'Build Own Operate' (BOO) basis. Whether the internationalization of the NTPC will be more successful than those of US and European utilities and whether other companies from developing countries will follow its example remains to be seen.

Alternatives to privatization or relying on foreign investment

While there is considerable discussion of the problems of publicly owned monopoly utilities, there is seldom any acknowledgement of the excellent track record many of them have over several decades in reliably expanding the supply of electricity so that a large proportion (or all) of the population have access to cheap reliable supplies. There is even less debate about whether the existing structure could be reformed without breaking it up so that the strengths of the old system – equity, accountability and development of local skills – could be retained whilst still addressing some of the perceived problems of the old system. The most commonly cited problems of the old monopoly model include:

- **Political interference.** Government ownership of the utility has often led to arbitrary and counterproductive political interference in the policies of the companies.
- **Lack of cost consciousness.** The monopoly status of the company leads to a lack of cost consciousness because costs incurred can be passed on to final consumers in a way that would not be possible in a market situation.

- **Inflexibility and exclusivity.** Centralized companies (public or private) are often too inflexible and may stifle to local or small scale options.

Fundamentally, the problems seem to arise because the relationship between company owners and management is badly managed. If a fraction of the effort that has gone into breaking up companies and trying to impose markets had been expended trying to reform the existing structure without breaking it up, the results might have been far better.

The World Trade Organization and the GATS negotiations

The Sri Lankan government is party to the General Agreement on Trade in Services (GATS) negotiations. The GATS provisions add to the pressure already being exerted by the IFIs such as the World Bank on developing country governments to break up their national electricity industries and open them to international investment. The liberalized and privatized model for the electricity industry is far from proven, especially in developing countries. Far from solving its problems, opening up the Sri Lankan electricity industry to international participation could therefore leave the Sri Lankan government with a system that does not work and which, under GATS rules, cannot be abandoned.

After the failure of the Cancún Ministerial Summit in September 2003, the GATS process appears to be on hold. The most powerful developing countries, including India, China and Brazil, led the formation of a loose association, known as the G21 (or the G22)[31] to press for a better deal for developing countries, particularly on agriculture. Another set, dubbed the G33, was a group of countries led by Indonesia and the Philippines and including Sri Lanka that formed the Alliance for Strategic Products and a Special Safeguard Mechanism which seeks special measures to protect vulnerable farmers. These, and various other developing country groupings, refused to agree to the demands of the developing countries and no agreement was reached at Cancún. However, negotiations continue and the GATS agenda has not been forgotten by the developed countries. Indeed, if the result of the Cancún failure is that powerful developed countries or developed country blocs like the USA and the EU start to emphasize more bilateral agreements, this could be very dangerous for developing countries, who could be 'picked off' one by one.

The prime focus for developing countries at Cancún and subsequently has been to obtain concessions on agriculture, and the GATS sectors may be drawn in to these as bargaining chips. While for some service sectors there may be a net gain for developing countries, for example by allowing concessions in service sectors in exchange for better access for food exports, it would be a mistake if these concessions related to the electricity sector. Opening up electricity could lead to a failure of that sector, the consequences of which would far outweigh any benefits from increased exports of agricultural products.

The World Trade Organization

History of international trade negotiations

The World Trade Organization (WTO) was established in 1995 as an international body to govern and expand international trade agreements (see Appendix 2 for sources of information on the WTO and the GATS). Its roots go back to the 1947 General Agreement on Trade and Tariffs (GATT), which was the first major international agreement aimed at reducing barriers, such as tariffs and dumping, to free trade. The WTO states that it does this by:

- administering trade agreements,
- acting as a forum for trade negotiations,
- settling trade disputes,
- reviewing national trade policies,
- assisting developing countries in trade policy issues, through technical assistance and training programmes, and
- cooperating with other international organizations.

It proceeds via a successive series of negotiations, or 'rounds'. One outcome of the Uruguay Round (1986–94) was the creation of the WTO, of which Sri Lanka was a founder member. There are now about 150 members of the WTO, accounting for about 97% of world trade, with 30 further countries negotiating membership. The European Union is a member and in most cases speaks for all 5 member states. The WTO's decision-making body is the Ministerial Conference, which meets at least once every two years. Below the Ministerial Council is the General Council (normally ambassadors and heads of delegation in Geneva, but sometimes officials sent from members' capitals) which meets several times a year in the Geneva headquarters. The General Council also meets as the Trade Policy Review Body and the Dispute Settlement Body. At the next level, the Goods Council, Services Council and Intellectual Property (TRIPS) Council report to the General Council.

The rationale for the GATT

The rationale for the GATT was that by removing trade barriers, production of individual goods and services would be exposed to competitive forces and consumers would be able to buy the cheapest goods regardless of their origin. This would mean that only efficient businesses would be able to survive and inefficient businesses that had previously survived because they were protected against competition in their home markets would have to emulate the efficiency of the best global companies. Countries with competitive advantages for the production of particular products or services, for example a particularly skilled workforce or advantages in natural resources, would be able to expand their markets. The lower prices resulting from closing down or improving the efficiency of companies would stimulate economic growth. According to economics theory, efficient markets lead to an optimal allocation of resources.

While on the face of it the basic logic of the GATT and free trade might appear indisputable, in practice there are circumstances where it is arguable that free trade policies are not appropriate. The following factors (by no means an exhaustive list) are of particular relevance to the electricity sector.

Barriers to entry. A central assumption of the free trade agenda is that there will be a field of international companies that will compete strongly in newly opened up national markets. Perhaps less obviously, there is an assumption that 'barriers to entry' are low. In other words, it will be easy for new companies to acquire capabilities and to enter markets. If barriers to entry are high, companies that fail or are taken over will tend not to be replaced by new entrants and the sector will become concentrated into an ever narrower field of suppliers. The clear risk in this situation is that market forces will be negated by monopoly or oligopoly power.

Strategic capabilities. In the past, governments have judged that the availability of particular products or services at stable prices was of crucial importance to their economy and that the international market in these products or services was not sufficiently reliable. In this situation, governments set up or protected local suppliers to guarantee the availability of this product or service. This logic has often been applied to commodities such as food and, for example, the European Union continues to have a Common Agricultural Policy aimed at guaranteeing the supply of food by protecting European producers. It has also been applied in the transport sector and the energy sector, particularly electricity, where publicly owned companies or heavily regulated private companies were the rule until the recent trend towards privatization and liberalization started to change this. Under the logic of the GATT, strategic national capabilities have little value because they prevent the development of efficient international markets, and because the companies are protected, there is insufficient pressure on them to be efficient.

Infant industries. Governments have, for reasons of national development, often targeted sectors in which they hope to develop internationally competitive national capabilities. It was argued that new companies or capabilities needed transitional protection so that they could build up their skills to internationally competitive levels. Such arguments were often applied in basic industries such as iron and steel. Particularly for developing countries, which often have quite a narrow industrial base, this was seen as a key way to broaden the industrial base and improve the international competitiveness of their economies. The European Union has also tried to encourage the development of European suppliers, for example in aerospace.

Preserving national capabilities. Particularly for highly cyclical products where periods of low demand might lead to the failure even of highly efficient companies, governments have often tried to protect national companies in periods of low demand or difficult trading conditions, perhaps by bringing forward orders or discriminating in favour of national suppliers. For example, the shipbuilding industry has often received support to prevent the closure of shipyards, while the British government has recently introduced strong support for a privatized nuclear power company to prevent its bankruptcy.

The principles of the GATT
The WTO identifies five main principles that underlie the GATT:[32]

- **Trade without discrimination.** This requires that countries apply 'Most Favoured Nation' (MFN) treatment to all signatories of the GATT. Under this, countries are required to treat all trading partners equally, or treat each country as well as they treat their most favoured nation partner. They are also required not to discriminate between national and imported goods and services, so-called 'national treatment'.
- **Freer trade.** This involves the progressive removal of all visible (e.g., import duties) and invisible barriers to trade.
- **Predictability.** Commitments made under the GATT are binding. Countries cannot re-erect trade barriers without paying

compensation to any country adversely affected by the change.
- **More competitive.** Under the GATT, unfair practices such as dumping and subsidizing exports are discouraged.
- **More beneficial to less developed countries.** A ministerial decision adopted at the end of the Uruguay round gave least-developed countries extra flexibility in implementing WTO agreements. It says better-off countries should accelerate implementing market access commitments on goods exported by the least-developed countries, and it seeks increased technical assistance for them.

These principles raise a number of issues, but of particular relevance to the electricity sector is the third, 'predictability'. This is seen as 'locking countries in' to a commitment, no matter how misconceived that commitment turns out to be.

The history of the GATS[33]

The Uruguay Round saw, for the first time, a major focus on trade in commodities other than goods, for example, services and intellectual property. The Uruguay Round culminated, in 1994, with the agreement of the General Agreement on Trade in Services (GATS), which came into force in January 1995. Under the GATS, the signatory member governments made a commitment to progressively liberalize trade in services. Article XIX (Paragraph 1) committed them to start a new round in 2000. A new round of negotiations was launched in 2001 in Doha under the banner the 'Doha Development Agenda', part of which was the GATS 2000 initiative. The Fifth WTO Ministerial Conference was held in Cancún, Mexico from 10th to 14th September, 2003.[34]

The GATS agreement has two parts. In the first, the general rules are set out, while in the second there are national 'schedules' which list individual countries' specific commitments on access to their domestic markets by foreign suppliers. Subsequent rounds of negotiations have concentrated on increasing the number of national services markets open to competition. These have produced 'protocols' with further commitments on financial services, basic telecommunication services and movement of natural persons. The WTO acknowledges that in the latter case little was achieved.

The GATS and the electricity sector

There has been extensive debate and opposition to the GATS in a number of sectors, especially water, where there has been a fierce debate about the extent to which the GATS will lead to water industries being taken over by multinational companies.

However, there has been very little specific analysis of its impact on the electricity sector. This is surprising, given that energy is such a major service both in terms of its economic significance and also its importance to consumers. One notable exception is the analysis by Cho and Dubash,[35] which argues that committing the electricity sector to international investment under the GATS would shrink the policy space open to developing country governments and jeopardize the ability of national governments to promote sustainable development in the electricity sector. They argue that governments will need to use policy instruments that might conflict with international investment disciplines. They cite a number of examples where the use of such instruments has been successful in meeting sustainable development goals (see Box 2).

Energy services were not negotiated as a separate sector during the Uruguay Round. The WTO reports that 'though a few WTO members undertook sparse commitments in various energy-related services the vast majority of the global energy services industry is not covered by specific commitments under the GATS'.[36]

Under the GATS agreement, 12 categories of service are identified:[37]

- business services,
- communication services,
- construction and related engineering services,
- distribution services,
- educational services,
- environmental services,
- financial services,
- tourism and travel-related services,
- recreational cultural and sporting services,
- transport services, and
- other services.

Within each of these broad categories, there are a number of subcategories. For example, within 'business services', there are six subdivisions:

- professional services,
- computer and related services,
- research and development services,
- real estate services,
- rental/leasing services without operators, and
- other business services.

These sub-categories are broken down even further and, for example, within the 'other business services' category, we find 20 further sectors, including 'services incidental to energy distribution'. This is the only specific mention of energy in the schedule of sectors. While 'services incidental to energy distribution' appears a rather limited set of activities, reference to a more detailed classification (the United National Provisional Central Product Classification) reveals that it includes core distribution and transmission activities. The issue of whether electricity generation itself comes under the scope of the GATS rather than under the general

Box 2: Examples of the use of non-market policy instruments in the electricity sector

- In Gabon, the government instituted a monopoly concession that bundled together the electricity and water sectors with incentives for service expansion, a policy that could conflict with GATS commitments on market access.
- In the 1930s, the United States government subsidized rural co-operatives to promote grid expansion, paving the way for universal electrification in the US, a policy that could now conceivably be interpreted as discriminatory under the GATS.
- The US state of Arizona provided competitive advantages to local solar manufacturers in the form of a performance requirement to guarantee local economic benefits from renewable energy.
- The Government of Denmark introduced a discriminatory tariff that privileged purchases of electricity from locally owned co-operatives, a policy inconsistent with the principle of national treatment.
- To mitigate a history of inequality, the Government of South Africa mandated ownership shares for black populations as part of public asset sales, and conditioned eligibility for government contracts on black ownership as part of a larger policy of 'Black Economic Empowerment'.
- The Government of Malaysia has conditioned industrial licenses on ethnic ownership guidelines, potentially diffusing political conflict among communal groups.
- To help resolve a financial crisis, the Government of Argentina imposed an electricity rate freeze and mandated renegotiation of utility contracts to spread the burden of crisis resolution to include foreign investors, a move that produced billions of dollars in claims for international arbitration under the terms of bilateral investment treaties.

provisions of the GATT is, as discussed later, unresolved.

How the GATS works

Under the GATS, there are four ways in which a service can be traded:

- **Cross-border supply.** Services are supplied from one country to another (e.g. international telephone calls).
- **Consumption abroad.** Consumers from one country make use of a service in another country (e.g. tourism).
- **Commercial presence.** A company from one country sets up subsidiaries or branches to provide services in another country (e.g. a bank from one country setting up operations in another country).
- **Movement of natural persons.** Individuals travelling from their own country to supply services in another (e.g. an actress or construction worker).

For electricity, it is clearly only the latter two provisions that are relevant and the third is the key one. The GATS is being implemented on a 'request offer' basis. Under this rule, participants had to submit initial requests for specific commitments by 30 June, 2002 and initial offers by 31 March, 2003. Requests are generally from one signatory to one or more other signatories, requesting an opening up of a specific sector. The request might be to add a new sector or to reduce restrictions on an already partly open sector. The process of 'requests' is a purely bilateral one between the requesting government and the target government, which the WTO is not usually informed about. Given the overall objective to open up trade in services, the WTO is keen to encourage requests. It states:[38]

'It is important to keep in mind that when each Participant submits an initial request it does not have to be exhaustive and a Participant does not necessarily have to think of every conceivable item it wishes to request of other participants. In order to meet the dates, it might be necessary to avoid seeking perfection which might cause delays.'

Requesting a market opening seems a largely cost-free exercise: essentially, if successful it opens up an option for companies from the requesting nation. If there is any likelihood that a nation's companies might benefit from the opening up of another country's market, there would seem to be no reason not to request the opening of a sector. A request commits the requesting country to nothing.

By contrast an offer is a major commitment. As with requests, offers may open up a new sector or reduce trade restrictions on a sector that is already partly open. This represents a major obligation, although the offer does not become binding until it has been the subject of negotiations and has been incorporated in the offering country's schedule of commitments.

Sri Lanka's initial offers made in the period 1995–9 are listed on the WTO web site.[39] No offers were made in the energy sector. The offers were in the telecoms, financial services and tourism sectors. If we look at offers in the energy sector made in the period 1995–9, the WTO database divides commitments on energy into four categories of countries:

- amongst developed countries only the USA and Australia made commitments and Australia's commitments only covered 'consultancy' services,

- amongst developing countries, only the Dominican Republic made a commitment on energy,
- amongst least-developed countries, no commitments were made, and
- amongst transition countries, Hungary, Kyrgyz Republic, Latvia and Slovenia made commitments.

The debate since 2000

Since 2000, the Council for Trade in Services has met in Special Session several times a year, but it was not until March 2003, that energy was mentioned as an explicit item on the agenda.[40] Energy did not seem to play a major role in the Cancún negotiations.

The initial lack of interest in energy services between 1995–9 led to pressure for greater commitments but since then only eight countries (or trading blocs) have submitted 'negotiating proposals'. In December 2000, the USA tabled a document intended to encourage discussion on services (Document 00-5556)[41] and to stimulate further commitments.[42] It defined 'services incidental to energy distribution' as broadly as possible to include 'exploration, development, extraction, production, generation, transportation, transmission, distribution, marketing, consumption, management and efficiency of energy, energy products, and fuels'. The document sought to explain the lack of interest shown in the opening up of the energy sector by, for example, the dominance in many countries of nationally owned monopoly suppliers, and identified barriers to the opening up of the sector, such as discriminatory regulatory regimes.

The USA proposed a separate and comprehensive section in the GATS on energy that would include all activities in the energy chain, including electricity generation as a service rather than as a good, which meant that generation should come under the provisions of the GATS rather than the normal goods provisions of the GATT.

The European Union followed this up in March 2001 with a further discussion paper on energy services (Document 01-1425). Like the US paper, it preached the virtues of liberalization and competition in the energy sector – 'a win–win opening up of national markets to competition and to foreign suppliers' – and exhorted signatories to the GATS to open up their electricity sectors. Like the USA, the EU took a very comprehensive view of the applicability of the GATS to electricity and energy in general.

Norway submitted a negotiating proposal in March 2001 (Document 01-1412) that covered several sectors including energy. Like the USA and the EU, it argued that energy production should fall under the auspices of the GATS and argued strongly for an opening up of the sector. Chile, in May 2001 (Document 01-2474), also in a document covering a number of sectors, supported the arguments put forward by the 'liberalizers', the USA, the EU and Norway.

Venezuela responded, also in March 2001, with a document that tried to promote the position of developing countries (Document 01-1552). It contradicted the US position that energy production should come under the GATS and stated that production should be seen as a good and come under the GATT. It also listed a number of provisions to protect developing countries. It stated:

- These negotiations should ensure that energy services are made

accessible to as many people as possible, in order to improve their standard of living, and to as many industries, businesses and services as possible, in order to promote economic growth.
- The market opening resulting from the negotiations should help to increase the energy supply capacities of all members.
- Furthermore, the agreements resulting from the negotiations should help developing countries to achieve improved access to technology and, in general terms, to pursue services-related policies designed to increase the competitiveness of all their production sectors.
- Consequently, the results of the negotiations should enable countries which use the supply of energy services as an instrument for boosting their development and as a means of diversifying their economy and strengthening the private sector to continue to pursue and to consolidate these policies.

More specifically, it stated:

- The negotiations should respect the appropriate flexibility for individual developing country members to open fewer sectors, liberalize fewer types of transactions and progressively extend market access in line with their development situation, in accordance with Article XIX of the GATS.
- The negotiations should respect the developing countries' space to implement policies aimed at domestic capacity-building, in particular the capacity of their small and medium-sized energy service suppliers.
- The ownership and rights of access to and use of natural resources are issues that should not be addressed in these negotiations.
- It did support the USA in calling for a separate energy classification to reflect the full range of activities in energy that were covered by the GATS.

A Japanese contribution in October 2001 (Document 01-4772) also supported the need for a specific classification of energy services and stated that energy production should not be seen as a service. The tone of this contribution was cautious to the opening of energy markets. It argued:

'When taking into consideration the following three points, i.e. countries with scarce energy resources; developing countries whose energy services industry is still at the developing stage; and the existence of differences in capital scale by country, energy security itself should continue to be of great importance in national energy policy. In particular, in light of the recent cases where liberalized markets have faced difficulties in supply, thus causing heavy burdens in a country in which regulatory reform of energy services is ongoing, it is also necessary to make efforts to ensure energy security and supply reliability when pursing regulatory reform and business reorganization.'

In March 2002, Cuba (Document 02-1500) broadly supported the Venezuelan position in its contribution, particularly stressing the need to take account of the special needs of developing countries.

The countries that had submitted proposals on energy services (USA, EU, Venezuela, Cuba, Norway, Chile and Japan) formed a self-styled 'energy friends group' that started meeting in 2001. After negotiations with the other seven members, in June 2003, Venezuela submitted a proposed detailed classification of energy services (Document 03-2883). It divided the sector into two parts, 'upstream' and 'downstream', with the downstream part of most relevance to electricity. Downstream services were split into three main sections:

- services for design, construction, and operation and maintenance of energy facilities, including networks,
- services for the commercialization of energy, and
- other energy services.

Within these three main categories were eight further divisions, for example, 'operation and maintenance of energy networks', and 27 further subdivisions, for example, within 'operation and maintenance of energy networks', there is 'operation and maintenance of electricity networks'. Production of energy was excluded from these proposals.

A constant theme through all the contributions has been the need for strong regulatory bodies, although in the case of those advocating liberalization, the objective is to ensure transparent and non-discriminatory rules, while in the case of those arguing a special case for developing countries, the objective is to allow countries to pursue domestic policy objectives.

Overall, it is clearly no coincidence that the strongest advocates of an opening of the energy sector, particularly electricity, are countries (or regions) such as the USA, the EU, Norway and Chile that have been amongst the most aggressive in opening their national markets to competition.

Progress on offers and requests

Very little information is published on commitments and none has been voluntarily published on requests to open sectors. The only substantive information on requests is the leaked schedule of requests made by the EU and published by the Canadian Polaris Institute.[43] The World Development Movement (WDM) has published a detailed analysis of these requests.[44] It shows the requests made, by sector, to 109 members of the WTO. Twelve sectors are listed, including energy.[45] In about 40% of the cases (46), the EU requested an opening of the energy sector. Compared to most other sectors, especially telecoms (106 out of 109), the number of requests to open up the sector was lower. Sri Lanka was requested to open all 12 sectors except energy and postal & courier services.[46]

There was a very poor response to the deadline for 'offers and requests'.[47] Only thirty countries submitted requests before the June 30, 2002 deadline. The only full requests in the public domain are the leaked EU documents. Only fifteen member states made offers before the deadline of March 31, 2003[48] and of these only five, Australia, Canada, New Zealand, Norway and the USA, have made their detailed offers public. The submissions of Australia,[49] Canada,[50] and New Zealand[51] did not include energy. Only the USA[52] and Norway[53] explicitly offered to open their energy sectors. The EU failed to reach agreement on its offers by the deadline and did not submit an offer until April 29, 2003.[54] There was no substantive reference to the energy sector in this document. The delay in reaching agreement between the member states of the European Union on what should go into the offer graphically illustrates that while making a request is a simple decision requiring no commitment, making an offer has far-reaching consequences.

General arguments on the GATS[55]

The general arguments for and against the GATS are complex and there is only space here to summarize some of the main ones.

Arguments for[56]

The main argument for the GATS is, as for free trade in goods, economic efficiency. By exposing services to international competition, they will become more efficient; consumers will pay less for these services and will therefore have more money to spend on other goods and services. Competition will also stimulate innovation.

The WTO argues that this process will be good for developing countries on a number of specific grounds. It suggests that service companies in developing countries will be able to take advantage of more open international markets to expand their businesses. It also suggests that the GATS will stimulate technology transfer. Foreign direct investment will bring with it new skills and technologies that local companies and employees can learn and adopt.

Arguments against[57]

The fundamental argument against the GATS is that the GATS agenda is being driven by the interests of large companies in developed countries. These companies will increasingly dominate world markets, not because of superior efficiency but because of greater resources, political influence and sheer market muscle. Without the opportunity to nurture local companies while they develop their capabilities, developing countries will be locked into a role as low-skill, low-wage providers of basic goods. Once commitments are made under the GATS to open a sector, government attempts to develop national capabilities by fostering local companies will be impossible and community-based non-profit service providers will find it impossible to survive against the might of multinational companies. Box 3 outlines the six main arguments against the GATS as put forward by the World Development Movement.

Key areas of dispute

The main areas of argument between the two sides are on:

- the extent to which the GATS will force the liberalization and privatization of vital public services, such as health and water,
- the reversibility of the GATS commitments, and
- the secrecy of the negotiations.

In the first area, the WTO argues that under the GATS, member states are free to retain existing arrangements, for example, a nationally owned monopoly service provider could be retained. Opponents argue that by signing up to the GATS countries have committed themselves to 'achieve a progressively higher level of liberalization in their service sectors'.[58]

'The Uruguay Round was only the beginning. The GATS requires more negotiations, the first to begin within five years. The goal is to take the liberalization process further by increasing the level of commitments in schedules.'

This, on top of the policies of the IFIs such as the World Bank that encourage privatization, puts heavy pressure on developing countries to liberalize/privatize services, regardless of the options technically open under the GATS. The developed countries such as the USA and the EU are also lobbying hard for privatization and liberalization.

The WTO argues that there are ways to reverse commitments. However, of the four mechanisms it lists, two allow only a temporary suspension of commitments, one is likely to require the payment of compensation and the other is applicable only in extreme circumstances such as endangerment of national security or public morals. If a government decided simply that a sector was more efficiently or reliably run as a public-sector monopoly rather than as a competitive private sector business, it is hard to see what grounds a country could invoke to withdraw a commitment. In other statements, the WTO is less optimistic about the possibility of withdrawal. In its 'introduction to the WTO', it states:[59]

> 'These commitments are "bound": like bound tariffs, they can only be modified or withdrawn after negotiations with affected countries – which would probably lead to compensation. Because "unbinding" is difficult, the commitments are virtually guaranteed conditions for foreign exporters and importers of services and investors in the sector to do business.'

On the charge of secrecy, the WTO counters that intergovernmental negotiations are inevitably secret, but that governments are the legitimate representatives of countries and that it goes to great lengths to publicize the results. The World Development Movement contrasts the involvement of industry lobbies in the development of proposals, particularly 'requests' to countries to open up sectors, with the minimal attempts to stimulate public debates on what sectors should be opened up in home markets and what governments should be asking, in the form of 'requests' of other countries. The heading on the 'request' to Sri Lanka from the European Union states 'Member states are requested to ensure that this text is not made

---

Box 3: The six key points from the WDM analysis

1. The EU is extensively targeting the world's poorest countries, demonstrating a massive imbalance in the negotiating capacity of rich and poor WTO members and providing a clear indication of who has most to gain from these talks.
2. The EU's sector-specific requests, if acceded to, will undermine countries' ability to regulate investment in the public interest. The EU's rhetoric about the 'right to regulate' cannot take away from the fact that the whole purpose of the GATS is to steadily remove the ability of governments to use their powers to direct investment in ways which benefit people, rather than companies.
3. The EU is seeking to remove a range of across-the-board regulatory rights in developing countries. The often repeated claims that GATS is flexible and that GATS does not undermine regulation are severely challenged by the exposure in the leaked documents of a list of regulations – specifically protected by developing countries in the last round of talks – which the EU is demanding be eliminated.
4. The EU is targeting countries where effective non-market-based service delivery systems are in operation. In contrast to its 'development round rhetoric' the EU's requests threaten the existence and further expansion of successful alternative forms of service supply (e.g. not for profit organizations and co-operative management systems).
5. The EU's requests target public services, despite its claims that it does not do so. The EU is clearly requesting GATS commitments in countries where the services in question are currently provided by the state.
6. The EU's requests threaten democratic policy making. The EU has demanded binding, effectively irreversible GATS commitments in the very countries where there has been popular resistance to – ultimately leading to government rejection of – certain liberalization policies.

Source: 'Whose development agenda? An analysis of the European Union's GATS requests of developing countries', by C. Joy and P. Hardstaff. World Development Movement, London, 2003: pp8–9.

publicly available and is treated as a restricted document'. This hardly suggests an open and democratically accountable process.

The application of the GATS to electricity

It is clear that energy has not been a high priority in the GATS negotiations so far, with few countries making any commitments in this sector. However, energy is one of the most economically significant services and is also one of the most vital services in a modern economy. There has also been a strong trend, particularly in the EU and the USA, to accelerate liberalization and deregulation of the electricity sector. Companies from the EU and the USA have been aggressive in moving into developing country energy markets. As a result, there is a strong movement driven by the USA and the EU, the two strongest trading blocs, to create a separate classification for energy services. This classification would be inclusive, bringing in parts of the electricity value chain such as electricity generation that are not self-evidently 'services'. The EU and the USA are also ideologically committed to a deregulated model for the electricity industry. In its proposal of December 2000, the US government states:[60]

'Competitive provision of energy services helps ensure that energy consumers have access to efficiently produced, market-priced, reliable energy. The availability of varied sources of energy at competitive prices, including access to supplies transmitted cross-border, contributes to a nation's ability to compete in the world marketplace. Competitive conditions in a nation's energy services markets enhance the competitiveness of domestic energy consumers as well as incentives for foreign investors to invest in both energy services and energy-consuming sectors. They also can benefit residential consumers and social services, as well as employment, through the beneficial impact on energy-dependent services and manufacturing sectors.'

While the EU in its proposal states:[61]

'The latest steps in the liberalization process have included the opening up to competition, subject to certain conditions, of the electricity and natural gas markets, which has resulted in significant price reductions for final consumers. The physical characteristics of these energy sources and the level of industrial concentration have led to the establishment of an appropriate regulatory framework, with the objective to avoid distortion of competition in the market. The process of liberalization of the energy markets has, in particular, created new opportunities for the supply under competitive conditions of a large range of energy services, some of which were previously carried out in-house by the monopoly companies.'

So it is likely that 'electricity services' will be of growing significance in future GATS negotiations and there will be pressure on countries not only to liberalize their existing market structure, but also to deregulate the electricity industry. Given the evident difficulty, if not impossibility, of reversing GATS commitments, this raises the question does the deregulated, liberalized model for the electricity industry work better than the centrally planned monopoly-based structure that it replaces?

The Cancún Summit and subsequent developments

There are many interpretations as to what happened at Cancún (see Appendix 6). However, with respect to the electricity industries of the developing countries, three points are clear:

- No substantive agreements were reached at the Cancún Summit and world trade negotiations are stalled.
- Developing countries showed far greater strength in negotiations than ever before, signalling that developed countries will have to offer much greater concessions than in the past if they are to reach agreement with developing countries.
- The GATS negotiations are not currently a major battleground in trade negotiations.

The immediate sticking point in the negotiations appears to have been the so-called 'Singapore Issues', trade and investment, trade and competition policy, transparency in government procurement, and trade facilitation. The Cancún Ministerial Conference ended on 14 September after Chairperson Luis Ernesto Derbez concluded that despite considerable movement in consultations, members remained entrenched, particularly on the 'Singapore' issues.[62]

Efforts began in October 2003 to re-launch the talks at an informal heads of delegation meeting at WTO headquarters in Geneva. Soon after, it was agreed that Hong Kong would host the next (6th) WTO Ministerial Conference, but no date was set. In November 2003, 12 African countries issued a statement calling for the resumption of trade talks. On December 15, 2003 at a WTO General Council meeting in Geneva, the WTO General Council Chairperson, Carlos Pérez del Castillo, stated:

> 'Members are willing to restart work in the negotiating groups, but there is still no major breakthrough.'[63]

By January 2004, no progress on the GATS appeared to have occurred since the Cancún Summit.

Conclusions

There are four main conclusions arising from our analysis of the Sri Lankan electricity sector:

1. The proposed reforms of the electricity industry will not solve the problems being experienced. New investors are unlikely to find the risks of investing in Sri Lanka to be acceptable unless strong government guarantees are given that would be potentially very costly and dangerous to Sri Lankan citizens. At present, even if the risks to investors were to be made tolerable, the retreat to their home territories of the international electricity utilities means investors would still be unlikely to emerge.

2. The new organizational model for the Sri Lankan electricity sector was put together for developed countries and is not appropriate for developing countries. The record of the British model in developed countries is at best mixed, and in the case of California, it was disastrous. In developing countries, it seems highly unlikely it will lead to the increase in investment and the improvements in efficiency that were expected, and in the case of Brazil, which did try to adopt the British model, its effects were almost disastrous.

3. The proposed reforms to the electricity industry could damage the prospects for decentralized power for isolated communities. For many Sri Lankans, the use of decentralized power sources represents the only realistic way for them to access electricity. Existing community-led systems have proved highly effective at meeting this need, but the proposed reforms could jeopardize existing systems and prevent development of new systems. For example, the new distribution companies will be given exclusive rights to supply electricity in their territory with no exemptions for isolated communities.

4. If Sri Lanka makes a GATS commitment to open its electricity sector to international investors, this commitment would be irreversible. This could mean that if the system proposed by the Sri Lankan government does not prove effective, it would be impossible for it to take the steps necessary to solve the problem.

The GATS

The GATS process is very one-sided, with the cards stacked in favour of the industrialized countries. Industrialized countries have far more resources and political power than developing countries and can ensure the process of 'offer and request' turns out favourably to them. The system of offer and request is unbalanced because industrialized countries, which have the resources to identify and target countries and service sectors where lucrative opportunities exist, can make requests that commit them to nothing and are merely 'door-opening' actions. By contrast, developing countries often do not have the resources to evaluate properly the consequences of what are essentially irreversible decisions to open up a market. Developing countries are under heavy pressure from industrialized country governments and from institutions such as the World Bank to open up and liberalize their markets. These may be seen as 'an offer they cannot refuse'.

The way in which the GATS is being negotiated is closed to the public and secretive. The process should be opened to the public so it can make an informed judgement on the commitments that countries like Sri Lanka make. Few of the offers and requests are being made public and, for example, the EU warns those receiving its requests not to publicize them. Such momentous changes to the way in which national economies are run should not take place without the informed consent of the public.

The Sri Lankan electricity system

The reforms to the Sri Lankan electricity industry endanger existing and future decentralized non-grid solutions for communities not connected to the central electricity grid. Decentralized power generation sources have proved themselves to be a reliable, cost-effective and environmentally desirable alternative to grid-supplied electricity for small isolated communities. Under the current Sri Lankan structure their legitimacy is not clear because the CEB and LECO have exclusive rights to distribute electricity. However, their advantages, acknowledged by government, and state ownership of the CEB and LECO, mean that their anomalous position has not been a problem. However, in a reformed electricity system, especially if ownership of the distribution sector is no longer public, it could be. It is therefore essential in any reforms that the position of existing schemes such as community-owned and operated integrated facilities are safeguarded and that isolated communities that decide to build their own decentralized generation and distribution system are not prevented from doing so.

The position of consumers without electricity that can reasonably be connected to the grid should not be forgotten: 25% of the population of Sri Lanka are in this position. Unless there are clear government commitments, and requirements on new investors where appropriate, extension of the grid to these consumers will not happen.

While the Ceylon Electricity Board may have failings, it has in the past proved itself capable of meeting growing electricity demand efficiently. Reforms to it that did not involve its break-up could mean that current problems could be solved, whilst still leaving intact an organization that could be a valuable tool in Sri Lanka's development. Liberalizing reforms around the world are running into a range of problems, from difficulty ensuring the right amount of investment takes place to unpredictable and rapidly rising electricity prices. The option of improving the CEB whilst still retaining its character as a publicly owned and accountable entity should be seriously investigated.

The widespread retreat of the electricity companies from markets outside the USA and Europe means it would be unwise to assume that opening up the Sri Lankan electricity sector would lead to investment by international companies. Any reforms that rely for new investment on foreign investors coming into the Sri Lankan market are very risky at present. International investors have lost large amounts of money in international markets and the major electricity companies are, at present, unlikely to want to invest in developing country markets because of the apparent risks. This will place countries that have reformed their electricity industry on the expectation of attracting foreign investment in a difficult position. They may be forced to shift even more of the risks of operating the system away from foreign investors and on to local consumers. For example, international investors will be wary about currency risk and risk of unstable demand patterns. They will also be unwilling to sign up to meaningful performance guarantees. This will leave local consumers bearing far more risk than they already do. Developing countries may also be forced to use companies that do not have sufficient financial and technical capability to the job properly.

The GATS and the Sri Lanka electricity system

It appears that the GATS negotiations are being used by industrialized countries to reinforce the already heavy pressure on developing countries to privatize and liberalize their electricity industries.

The liberalized, privatized electricity model is clearly unproven in developing country conditions. If energy – and specifically electricity – is to continue to be part of the GATS agenda, industrialized countries should be far more cautious in recommending an unproven model to countries that often do not have the resources to judge whether such a model is appropriate. Similarly, developing countries should resist pressure to open up their electricity industries until there is strong evidence that the new model really will meet the needs of their population, especially the poorest people.

Appendix 1: Retreat of multinational electricity companies

A. US Companies

There has been a massive focus on the activities of Enron, particularly following its collapse in 2001. In many of the markets it entered, there was widespread dissatisfaction with its activities.[64] Despite its bankruptcy in 2001, its foreign investments are only slowly being divested. For example, in July 2003, it finally sold its stake in a Puerto Rican utility, and in January 2004, it still controlled Elektro, an electricity distribution company in Sao Paulo, Brazil that supplies 1.7m consumers.[65] However, Enron was only one of a number of US utilities that followed a policy of expansion into international electricity markets.

AES

AES (Applied Energy Services) is an important company for Sri Lanka as it already owns the 168 MW combined cycle plant at Kelanitissa that was completed in 2003 and is the type of company expected to invest in Sri Lanka. AES is a very different company to the other US companies that have been active in the international electricity industry. Unlike the other US companies which (with the exception of Enron) were based around old, traditional US electricity utilities with a franchise territory, AES is a relatively new company, formed in 1981, whose business grew around the construction and operation of independent power plants (IPPs). Whilst a number of its international ventures are in serious difficulties, unlike the traditional US utilities it does not have a stable US base it can retreat to. Nevertheless, since mid-2002, it has been selling its non-US assets to raise cash to overcome financial problems.

AES began to look outside the USA in 1989 and its first major acquisitions were in 1992 in Northern Ireland and Argentina. Its business continued to be dominated by IPPs or other generation interests until 1996, when it began a concerted move between 1996–2000 into Latin America, with major utility acquisitions in Brazil, Argentina, El Salvador, Dominican Republic and Venezuela as well as generating plants in Chile, Panama and Mexico.

From 1999 onwards, its focus seemed to change to the former Soviet Union for utilities and for generating plant to Africa and the Indian sub-continent. Its largest purchase was the 4000 MW coal-fired Drax power station in England in 2000, for which it paid US$3 billion.

Its operations were frequently controversial, but up until 2001 its record of profitability was consistently good. Profitability began to decline steeply then and in 2002 it posted losses of US$3.5 billion, of which US$3066 million was impairment charges [impairment charge is a term for writing off worthless goodwill]. This was primarily due to problems in two markets, the UK and Brazil. Asset impairment charges in Brazil and UK amounted to US$2.3 billion. Other charges included US$301 million for businesses that have been sold or closed down, which included assets in the USA, Australia and US$465 million on certain development projects.

In their 2002 accounts, AES took asset impairment charges of US$1 billion in relation to its UK assets. In August 2003, after accumulating debts of nearly US$2 billion, AES effectively walked away from Drax, which was repossessed by the banks that had underwritten the loans. Drax, the largest coal-fired plant in Europe, still remained in the hands of creditors in January 2004.

In Brazil, it had a much larger range of facilities. It controls the largest distribution company in Sao Paulo (Eletropaulo), a distribution company in Porto Alegre (AES Sul) and has a minority stake in a large integrated company in Minas Gerais (CEMIG). In generation, it controls a number of plants including the huge Tiete hydro-electric plant (2600 MW) and a large gas-fired plant (Uruguaiana, 600 MW). Its problems stemmed partly from the reduction in value of Brazil's currency (the real) compared to the US dollar and also to the power shortages in Brazil in 2001. This resulted in charges of US$706 million relating to the Eletropaulo distribution company and US$587 million on other assets. The Brazilian nationally owned Brazilian Development Bank, BNDES, lent AES the money to finance the purchase of Eletropaulo. By the end of 2002, the outgoing Cardoso government had granted AES several extensions on repayment of loans totalling US$1.2 billion to BNDES. The new government was tougher on these extensions. After lengthy negotiations, in December 2003 a deal was reached between the government and AES which would see US$600 million of the debt written off. Under this deal, AES's stakes in Eletropaulo, AES Tiete and AES Uruguaiana would be transferred to a new company, Brasiliana Energia. AES would own 50.01 of the common shares and 53.85% of the voting shares with BNDES holding the balance. In addition, AES would pay BNDES US$90 million on completion of the deal and the remaining US$510 million (plus interest) over 11 years.

There is also an ongoing investigation in Brazil into whether Enron and AES colluded over the sale of distribution companies in Sao Paulo, including allegations that Enron had chosen not to bid for Eletropaulo in exchange for contracts with AES, allowing AES to buy the utility at the government's minimum price.

Write-offs in Argentina, which like Brazil experienced a major devaluation of its currency, amounted to US$190 million, although the company warned of further write-offs in Argentina in 2003.

In 2003, AES carried out a number of asset sales and withdrew from six of the 33 countries it was operating in. In February it sold a US utility (Cilcorp) for US$1.4 billion. In March, it announced the sale of power plants in Bangladesh, a stake in a subsidiary with power plants in Oman, Qatar, and Pakistan and its stake in two power plants in California with total proceeds of US$327 million. In April it announced the sale of two power plants in South Africa and Tanzania, also for US$327 million. In August, it allowed its Drax holding to be repossessed and abandoned construction of a large hydro-electric plant (Bujagali) in Uganda, forcing the Ugandan government to seize all AES's assets in Uganda. In October, it sold a UK power plant for US$78 million.

Southern Company/Mirant

The Southern Company was based around five US electricity utilities in four states: Alabama Power, Georgia Power, Gulf Power, Mississippi Power and Savannah Electric. By 2003, these companies owned about 37 GW of generating capacity and Southern Company was one of the largest generators in the USA. In 1981, it was given unprecedented approval by the Securities and Exchange Commission (SEC) to set up an unregulated subsidiary, Southern Electricity Inc. (SEI) (later renamed Mirant).

It was the first of the traditional US utilities to try to expand outside the USA, buying the South West Electricity Board (SWEB) in the UK in June 1995, taking a share in the Berlin utility, BEWAG, in September 1997, and buying a majority share in the Chinese company, Consolidated Electric Power Asia (CEPA), in October 1996. It was also a major investor in Brazil through its acquisition of a small stake in a large integrated electricity company, CEMIG.

In March 2001, Southern Company floated off Mirant as an entirely separate company but it soon began to sell its remaining assets outside the USA, including its remaining stakes in SWEB and BEWAG. The Southern Company remains a US-only company.

It began to withdraw from some of its South American markets after making heavy losses on them (US$200 million), putting its stakes in the Alicura hydro-electric plant in Argentina and the Edelnor utility in Chile up for sale in May 1999, overtly to concentrate on Brazil through its joint share with AES in CEMIG. However, in September of that year, the governor of Minas Gerais began proceedings against AES/SEI to remove their voting rights in CEMIG. The case went through various courts and its stake was eventually sold to an undisclosed buyer in January 2003 for a nominal amount. The Edelnor business was sold in 2002, indirectly, to the Belgium/France-based Tractebel company and the Alicura stake to AES.

However, SEI's most humiliating reverse was with BEWAG. In 1999, mergers in the rest of the German electricity industry raised the prospect that its German partners, VEBA and VIAG would have to sell their stakes in BEWAG if their merger was to be allowed by the German competition authorities. SEI hoped to take over BEWAG and also the former East German company, VEAG, to form one company to rival the three other major players in Germany. However, after a protracted struggle, in which SEI raised its stake to 44.8%, the Swedish publicly owned company, Vattenfall, which already owned the Hamburg utility, took over BEWAG and Mirant sold its stake to Vattenfall in December 2001 at a profit of US$1.6 billion.

While this was happening, major changes in SEI were taking place. In April 2000, the Southern Company announced its intention to spin off SEI as a separate company, subsequently renamed Mirant, and in April 2001, Mirant was launched as an entirely separate company valued at US$9.4 billion.

It continued to try to develop its European business, buying a stake in a planned Norwegian gas-fired power plant. Mirant was initially highly successful and its shares reached US$47 in its first year, but by 2003, they were trading at little more than US$1 and Mirant filed for bankruptcy on July 14, 2003.

PPL

PPL Corporation, originally Pennsylvania Power and Light, was a traditional electricity utility founded in 1920 serving Central Eastern Pennsylvania. In 1994, Pennsylvania Power & Light formed a holding company, PP&L Resources, Inc., to serve as parent to the regulated electricity utility, and to a newly formed, unregulated subsidiary called Power Markets Development Corp., created to invest in power projects domestically and overseas. Power Markets Development Corp. later became PPL Global. Its first major foreign investment (in 1996) was the purchase of a 25% interest in a British regional electric company, South Western Electricity, plc, from its owner SEI. It subsequently increased its holding to 51% (1998) and 100% (2002). It sold the retail part of the business in 1999 (to EDF) retaining only the distribution business (as Western Power Distributors). It bought another British distribution business (Swalec) in 2000.

In 1997, it purchased a 25% stake in a Chilean company, Empresas Emel, which had interests in Chile and Bolivia, and in 1998, Emel bought a majority share in an electricity distribution business in El Salvador. In 1999, it increased its stake in Emel to 95%. In June 2000, it purchased an 85% share in a Brazilian distribution company, CEMAR. In August 2002, PPL declared CEMAR bankrupt and it was repossessed by the Brazilian development bank, BNDES, the company that had financed its purchase.

TXU

TU Electric was formed in 1984 from the merger of three Texas companies and became TXU in 1999. In 1995, TU went international with the purchase of an Australian electricity distributor, Eastern Energy. In 1998, its first investment in Europe was the takeover of a British distribution company, Eastern Electric, for £4.4 billion (US$8.3 billion). In 1999, it began to expand into mainland Europe partly as an electricity trader, but also taking over a German company, Stadtwerke Kiel. It also took over another regional retail company in the UK, Norweb. The European businesses (including UK) were known as TXU Europe. It tried to establish an aggressive trading business across Europe in gas and electricity, emulating Enron's example.

In 2001, it began to sell fixed assets, such as the network of Eastern Electricity (UK) and some UK power plants. In October 2002, it sold most of its UK business to the German company, E.ON, and TXU Europe was declared bankrupt, leaving TXU with its businesses in Texas and Australia.

PSEG

Public Service Enterprise Group is a long-standing electricity utility based in New Jersey, USA. In 1990, Community Energy Alternatives, later renamed PSEG Global, was created as a subsidiary to develop independent power projects worldwide. In 1998, it made its first major utility purchase, buying companies in Brazil (Rio Grande Energia) and Argentina, although subsequently it sold its Argentine assets to AES in 2001. It set up PSEG India which was to be a 'hub' for investments in India, Sri Lanka, Bangladesh, Nepal and Oman. Its Americas division still owns distribution companies in Chile and Peru as well as small power plants in Peru and Venezuela. Its European division owns small plants in Poland, Tunisia and Italy, while in Asia it owns power plants in China, India (one plant, the only operating plant now owned by PSEG India) and Taiwan. Its total global generating capacity in IPPs is only 2500 MW.

Entergy

Entergy was based on a multi-state US electricity company called Mid-South Utilities, which became Entergy in 1989 and which took over another multi-state company, Gulf State Utilities, in 1992. In the mid-1990s, it bought electricity utilities in the UK,

Australia, Argentina, Chile, China, Pakistan and Peru. However, after poor financial results in 1997 and 1998, Entergy sold nearly all its non-US assets and now concentrates solely on its traditional US markets.

Others

Other companies, such as Reliant, Mission Edison and NRG have invested outside the USA but none are still expanding and some are withdrawing from their existing foreign markets.

B. European companies

The largest electricity utilities in the world are nearly all European or Japanese. However, the Japanese utilities (TEPCO and Kansai Power are the largest) have so far shown no interest in acquiring companies outside their home territories. Amongst the largest European utilities, the German companies (RWE and E.ON are the largest), Vattenfall (Sweden), and ENEL (Italy) own little outside Europe. The main global companies are EDF (France), Endesa and Iberdrola (Spain), and Tractebel (the energy division of the French company Suez).

EDF

EDF was one of the first European utilities to invest outside its home market, moving strongly into Latin America and taking over distribution companies in Argentina (Edenor, 1992) and Brazil (Light, 1996). It also has small IPP interests in Mexico, Cote d'Ivoire, Morocco and China. However, its main expansion has been in Europe, where it is a strong player in countries such as the UK, Germany, Spain, Italy, Sweden, Austria, Hungary, Poland, and Switzerland. Its assets outside France and the UK are now doing poorly and in July 2003 there were strong rumours that it was trying to sell its Brazilian business. The proposal by the Chirac government to privatize the company may mean it will abandon further attempts to expand outside France.

Endesa

Endesa of Spain acquired a number of Latin American electric companies through its takeover of the (till then separate) Endesa de Chile in 1999. Endesa de Chile was the dominant company in Chile and also had investments in Argentina, Brazil, Peru and Colombia. It has no significant businesses outside Europe and Latin America.

Tractebel

The French multinational company, Suez, has major energy holdings through its subsidiary, Electrabel, which dominates the Belgian electricity industry and through Tractebel, a company controlled by Suez, which owns electricity assets worldwide. Outside Europe, Tractebel's main markets are in North and South America and South-East Asia. It owns no significant assets on the Indian sub-continent.

German companies

The German market is dominated by just two companies, RWE and E.ON, which between them control about 80% of the German market. Both companies' major foreign investments are in the UK and Eastern Europe, for example in Hungary, Czech Republic and the Slovak Republic. In the UK, RWE took over Innogy (the daughter company of National Power) and E.ON took over Powergen, so German companies now represent two of the five main companies in Britain. Both E.ON and RWE expressed interest in 2002 in expanding into North America, but both appear to have abandoned this objective and are now concentrating solely on European markets.

British companies

It might be expected that as pioneers of liberalization, British companies would have an advantage in world markets, but of the 18 companies privatized, all except four have been taken over or broken up, including the two large generation companies, National Power and Powergen. Of the four remaining companies: British Energy, the nuclear generation company, is near bankrupt; National Grid Company specializes only in high-voltage transmission; Scottish Power's only foreign investments are in the USA; and Scottish & Southern's policy is to operate solely in the UK.

Appendix 2: Information on the WTO and the GATS

The World Trade Organization website (www.wto.org) contains a wide range of promotional materials on the organization.

The World Development Movement website (www.wdm.org.uk) provides critical analysis of the GATS process both from in-house produced materials and links to other sites.

GATSwatch (www.gatswatch.org) is a joint project of Corporate Europe Observatory (www.corporateeurope.org) and Transnational Institute (www.tri.org) that monitors the negotiations and provides critical analysis.

The European Union's information point on GATS used to be http:\\gats-info.eu.int/gats-info/gatscomm.pl, but in January 2004 this link no longer worked.

The World Resources Institute www.wri.org/wri/index.html

The Foundation for International Environmental Law and Development (FIELD) www.field.org.uk/index.php

Appendix 3: Perceptions of the Sri Lankan electricity industry

A. The planning process

Recent interviews with stakeholders[66] elicited the following perceptions about the planning process:

- No overall national-level policy to guide the sector.
- Political, commercial and sometimes improper financial considerations are alleged to influence decisions.
- There are presently several different mechanisms available for the development and approval of power projects, hence a proliferation of committees and authorities, which have led to a proliferation of points of views and lack of finality to the process.
- No clear indication as to who takes the final decision, hence a lack of certainty and accountability.
- CEB's long-term generation plan was not implemented as planned.
- Crisis purchases have lead to circumventing of regular procedures.
- No clear structure or legal mechanism to handle public dissent.
- Lack of co-ordination among key players despite the fact that the same people are often involved in the various decision-making bodies.
- Lack of integrated planning means that decisions on power supply systems do not look into other related areas that use energy such as irrigation, water supply and transport.
- Tied loans and other conditions laid down by donor agencies result in policy decisions that are not always in the best national interest.
- Stakeholders' opinions tend to be heavily polarized, therefore compromise is difficult to achieve.

B. Consumer perceptions of the problems

The consumer has had to bear the burden of this crisis in terms of increased prices, cost of living and inconvenience. Even those who do not have electricity have borne the cost of the power shortage. In order to determine opinion of domestic consumers and public awareness regarding issues relating to the power crisis surveys were carried out over a period of three weeks in early 2003.[67] Over 250 surveys were carried out and opinions were gathered from Colombo, Battaramulla, Moratuwa, Panadura, Kiribathgoda, Gampaha, Homagama, and Pannipitiya. Areas out of Colombo such as Kandy, Galle, Hikkaduwa, Dambulla, Batticaloa, Nikawaratiya, Anuradhapura and Puttalam were also reached. The people surveyed were divided into user groups based on their monthly expenditure on electricity in order to determine if opinions varied significantly among user groups. The majority of those surveyed were medium users (53%), 31% were low users and 16% were high users of electricity. Most spent between Rs1000–3000/month on electricity. The national grid supplied 93% of those surveyed, while 7% used other sources of power, mainly batteries, kerosene and solar.[68]

The main results were:

- Of those surveyed 71% felt that the price of electricity was unreasonable. This view remained consistent among the different user groups.
- When asked to categorize the service they received as 'good' 'moderate' or 'bad', most felt the service was moderate. This view was consistent across all user groups, irrespective of whether the supplier was CEB or LECO.

- Almost equal percentages of people were unwilling (38%) or could not afford (37%) to pay more for a better service. However 33% of the medium user group was willing to pay more, as opposed to 12% of the high user group and 18% of the low user group. In addition 51% of the lower user group said they could not afford to pay more.
- In terms of privatization, 45% of the high user group and 40% of medium user group felt that privatization would provide them with a better service, while amongst the low user group an equal number (39%) felt it would not or did not know if it would. A majority expected price increases to accompany privatization.
- The main reasons for the recent power cuts were ranked as mismanagement, bad planning and poor rainfall respectively but with little difference in percentages (27%, 24% and 21%). People also felt that wastage of power (lighting up of billboards and streets during festival periods) and an over dependence on hydropower generation were contributing factors to the crisis.
- Consumers did not favour an increase in prices or the cutting of power when the country was facing a power shortage, but advocated other methods such as reducing power wastage by fining wastage, introducing incentive schemes for reduced usage, or using energy-saving devices. Suggestions were also made for decentralized power generation and more emphasis on renewable sources.
- The effects of the power cuts for households included increased expenditure on measures to alleviate the shortages, e.g. energy-saving bulbs, generators and inverters.
- Those surveyed felt that Sri Lanka should invest in renewable sources to meet future power requirements. Solar was ranked first, with small hydro ranking second and wind energy third. Solar was also ranked the most economical option as the energy source was seen as being free. This shows that the public would like to see cleaner energy options being adopted but it also can be argued that the technical details of a power generation system are not clearly understood by the public.
- Some 68% of those surveyed were willing to pay more for an environmentally friendly source of power. However this figure dropped to 55% in the high user group, and increased to 72% in the medium user group.
- Most considered coal to be the most environmentally damaging source of power, with diesel ranking second. No opinion was expressed by 7% of those surveyed.
- Solar power is considered to be the least environmentally damaging source of power by 42% of those surveyed. Wind and small hydro ranked second and third respectively.
- A majority of those surveyed felt that media coverage gave them a moderate understanding of the issues and expressed interest in accessing more information on issues such as technology, management and costs.

C. The reforms

The following is a summary of concerns raised in 2003 after a joint study of the proposed reforms by

ITDG, Energy Forum and the Citizens' Trust.[69]

- The reforms tend to focus only on the main grid electricity system, whereas over recent years considerable progress has been made towards the establishment of an environment and a market for decentralized energy supply (micro-hydro, solar, wind and bio-fuels) to supply the needs of poor and remote communities. There is concern that the reforms and current policies will have negative effects on the progress of these decentralized, community-based schemes, unless mechanisms are set in place to address them.
- There appears to be no national energy policy for Sri Lanka. The two documents setting out, respectively, power sector policy guidelines and a policy for rural electrification, post date the gazetting of the two key power sector reform laws, namely the Electricity Reforms Act and the Public Utilities Commission Act. This begs the question as to what extent there is actual policy formulation based on scientific and socio-economic study, or whether 'policy' in this context is merely a rationalization of existing law.
- The published rural electrification policy does not take into account the power sector reforms and the implications it would have. Furthermore, although there are policies, there are no strategies for rural electrification, especially the 20% of households that cannot be connected to the national grid.
- It is important at the outset to incorporate suitable measures to ensure the status of these off-grid systems in the new reformed power sector.
- The off-grid community-owned electricity generation systems do not have a legal status as only CEB is currently mandated to generate and distribute electricity. These systems however are recognized and promoted by the national and provincial governments.
- The proposed reforms prevent a single license holder from both generating and distributing electricity.[70] Under the Electricity Reform Act the PUC may grant exemptions from the licensing requirements by Order published in the *Gazette*, having regard to the manner in which, or the quantity of electricity likely to be generated.[71] However the legal status of such schemes and their relationship to the Regulator is unclear.
- The proposal to give one distribution company exclusive distribution rights to a geographical area will have serious implications on community-owned systems. The legal status of these systems needs to be clarified so as to ensure that decentralized distribution systems outside the grid will be allowed and enjoy legal protection.
- Electricity is a devolved subject.[72] Currently, the Provincial Councils and local authorities play an important role in rural electrification. There are significant numbers of off-grid systems that are promoted by Provincial Councils, especially in Uva, Sabaragamuwa and the South. The policy guidelines do not refer to this link. The relationship of the Public Utilities Commission to the Provincial Councils needs to be clarified.[73]
- Grid-connected mini-hydro schemes use the water resources that belong to the surrounding

communities, who directly and indirectly conserve and manage these resources. There is an unethical situation created by mini-hydro projects using community-owned resources without giving back any benefit to the community (income or electricity connections). Provision should be made for some part of the profits generated from these schemes to be invested in the economic and social development of the communities surrounding these sites.
- At present there appears to be little co-ordination between the various government departments and donors who are planning initiatives in energy, rural energy, and renewable energy and the movers of the reform process. This needs to be addressed.[74]

D. Rural electrification

Issues raised in this sector include:[75]

- While it has been a priority area for successive governments, rural electrification has been done in an ad-hoc manner, largely dependent on availability of external funding. No clear plans have been made.
- Rural electrification by grid has resulted in losses for CEB as grid connection to remote areas is expensive and results in greater transmission losses.
- As the power sector suffers from a deficit of supply and the medium-term commitment is to reduce the deficit and strengthen existing systems, rural electrification may get put on hold.
- The poor pay more per kWh for electricity if it is not from the national grid, as they have to pay for installation and technology as well as consumption.
- Difficult political decisions need to be made as to which areas should get priority.

As power becomes accessible consumption patterns change and eventually the power sources/systems available must be able to meet the demand. Rural communities have been contributing towards increasing the use of renewable energy sources that have both provided them with energy and had overall benefits to the country in terms of a reduction in emissions. This is a cost/benefit factor not generally taken into account.

It has also been argued that energy delivery for the rural sector should address better overall energy needs whether it be for cooking, lighting, heating, or rural industries.

Appendix 4: The CEB generation plan: 2002–16

Power projects have long life spans. The power sector receives a large share of national investment and decisions taken on power sources affect other sectors. Therefore a long-term generation plan is needed. The task of preparing this plan has to date fallen on the CEB.

This plan needs to specify the technologies, fuel options, capacities of power plants and timing. The transmission and distribution plans are based on the power projects that are chosen. Therefore the generation plan becomes a key element of power sector planning. The CEB generation plan is derived using a computerized mathematical package (WASP)[76] that picks the best option through a cost optimization[77] process aimed at providing a reliable electricity supply at the lowest possible cost.[78]

The steps are:

- developing a long-term load forecast for a 20-year period using amount of power consumed by sector (domestic, industrial, etc.) and consumption patterns throughout the day (i.e. base load, peak load),
- study existing power plants,
- screen new technology and candidates for new power plants,
- study fuel costs, investment costs, etc. along with their efficiencies for the type of function they will provide – base load, intermediate load or peak load,
- feed the specifics (costs, capacities, consumption, maintenance costs, etc.) of different types of power plants into the system as input parameters,
- allow the model to select the best combination of power plants and time to introduce them,
- check options for robustness and sensitivity to different demand forecasts and adjustments are made, and
- review and revise this plan annually.

Hydro power plants are decided on separately and fed into the plan. If hydro power plants get funding, as in the case of Upper Kotmale, they are forced into the plan. Hydro plants have a long life (over 50 years) while thermal plants have a much shorter life-span (typically 35 years). Using a 10% real discount rate[79] the high construction costs of the hydro plants make them appear more expensive than thermal options and they are generally not chosen.

The generation plan is one of the most technically comprehensive and long-term documents produced by a public institution in Sri Lanka, drawing together information collected from various studies over a long period of time. According to CEB sources, the plan is discussed periodically and approved by CEB management, which includes representation from generation projects and transmission and distribution managers. Nevertheless there are criticisms, mainly that:[80]

- it is formulated in isolation, without transparency or public or inter-sectoral dialogue, with no opportunity for independent verification of stated facts including costing,
- it is a highly technical document that might not be easily understood to lay persons or professionals outside the sector and therefore difficult to give valuable inputs. An indicative plan if developed would be able to be discussed more widely,
- the choices picked by the models depend on the variables put in, and these can be manipulated,

- social and environmental costs are not included or given enough weight., and
- the planning methodology is developed for centrally planned vertically integrated utilities.

This makes no allowance for the proposed reforms.[81]

Table 1 shows the Power Plants proposed under CEB Generation Plan (2002–16).

Table 1: Power Plants proposed under CEB Generation Plan (2002–16)

	Capacity (MW)	Forecast completion date	Year completed	Remarks
THERMAL – GAS/OIL				
Ace Power – Matara	20	2002	2002	Operational on a BOO basis with a power purchase agreement with the CEB (accounted for in Table 1).
AES – CCY Gas (Phase 1)	109	2002	Complete	ADB Funded. A BOT project with a 20-year power purchase agreement. Phase 1 was operational but shut down for the present to install Phase 2 and due to be online soon. (Phase 1 accounted for in Table 1.)
AES – Steam – CCY (Phase 2)	54	2003	Not complete	
JBIC – CCY (Phase 2)	61	2003	Not complete	To be CEB owned and operated. JBIC has funded a CCY project and Phase 1 for 104 MW is operational.
Ace power – Horana	20	2003	2003	Completed original site in Anuradhapura but moved to Horana due to public protest. Ready ahead of schedule in Dec. 2002 but power purchased only in 2003 due to prior purchasing of emergency power. On a BOO basis.
Peilstick diesel – Sapu	22	2003		CEB owned and operated.
Kerawalapitiya CCY (2 × 150)	300	2005		EOI has been issued and pre-qualifed bidders have been identified. This is intended to be on a BOT basis.
Gas turbine	105	2007		No information was found regarding this.
THERMAL – COAL				
Coal1 – West Coast	300	2008		EOI issued by ESC in Nov 2002, and amended in 2003, restricting the fuel source to coal and the amount of power to 300 MW. Sites in Trincomalee or Hambatota, to be decided by the ESC. On a BOT basis.
Coal2 – West Coast	300	2010		
Coal3 – West Coast	300	2012		
HYDRO				
Kukule	70	2004		Originally scheduled for 1999, commissioned 2003. Funded by JBIC and owned and operated by CEB.
Upper Kotmale	150	2008		Originally scheduled for 2000, multiple delays. Funded by JBIC and owned and operated by CEB. Controversial.

	Capacity (MW)	Forecast completion date	Year completed	Remarks

CEB has also proposed the following sites as possible hydro projects:

	Capacity (MW)	Forecast completion date	Year completed	Remarks
Uma Oya	150			Studies done.
Gin Ganga	49			A run-of-the-river project, close to the border of the Sinharaja World Heritage site. Controversial location.
Broadlands	40			A run-off-the-river project, close to the Polpitiya power plant. Some discussions seem to be taking place.
Moragolla	27			

NEW ADDITIONS – NOT MENTIONED IN THE CEB GENERATION PLAN[82]

	Capacity (MW)	Forecast completion date	Year completed	Remarks
Thermal – Diesel	200	2003		Tender process ongoing. Called for by the ESC on authority from the Prime Minister.
Renewables	85			World Bank RERED project, to produce 5% of the total electricity generated from privately owned grid connected renewable sources. Additional provision for 100 000 households and 1000 SMEs/public institutions in rural areas to get electricity through stand-alone or village-owned renewable energy systems.
Wind	20			Tenders called for private generation of a plant to be located either at Puttalam or Hambatota.
Micro-hydro	41			Mahaweli Authority has called for EOIs for 18 locations. BOO basis. Closing date for applications February 3, 2003.

Appendix 5: Examples of operating micro-hydro projects

Kithulritiella village micro-hydro project, Perupalla, Maliboda, Deraniyagala

Project Details

A hydro project completed in the year 2001 with a capacity of 5.5 kW provides electricity to 16 households as well as to the temple in the village. The total cost of the project was Rs728 000. A loan of Rs240 000, repayable over five years, was obtained by the beneficiaries from the DFCC Bank under the World Bank ESD project for the project construction work and its repayment is near completion. An Association of Electricity Consumers formed by the 16 beneficiary households – the Kitul Riti Ella Electricity Consumers' Association – is responsible for the management, operation and maintenance of the hydro project. Each electricity-using household pays a monthly fee of Rs150 for the electricity that they consume and another Rs180 as their monthly loan installment. The membership fee of the Consumer Association is Rs20. The total number of beneficiaries from the village hydro scheme is 16 families, plus the village temple. Three other families are to be provided with electricity.

Entrepreneurs using electricity at present

1. **Jayasinghe** – Battery charging. Has capacity for five batteries 12v, 90 amp-hours. 6–7 batteries are charged per month. Rs60 charged per battery. Electricity used for 12 hours.
2. **Senaratne** – Battery charging. Has capacity for 10 batteries. 6–7 batteries charged per month. Rs50 charged per battery. Electricity used for 12 hours.
3. **Jayantha Kumara** – Radio-repairing, and uses a booth for 1 hour a month. He needs further training and daytime electricity to improve his enterprise. He now repairs five radios a month on average.
4. **Ms. Rushan Ranasinghe** – She does sewing from a rented facility (temple premises). She runs a juki machine and an over-lock machine for about 1.5 hours per day. She hopes to buy another machine and expand the enterprise with her son who does tailoring in Colombo. She also sells ready-made garments and earns about Rs3000 per month. She has no problem finding a market for her products.

Thanthrikanda village hydro project, Thanthrikanda, Miyanawita, Deraniyagala

1. Generation Capacity: 7.5 kW
2. Total project cost: Rs946 000
   – Grants from Provincial council: Rs255 000
   – Loan from ESD project: Rs300 000
   – Community Contribution: Rs175 000
3. Number of households – 51, School – 1, Temple – 1
4. Rural economic activities – Battery charging centre: 1, Small carpentry workshop: 1
5. Monthly instalment: Rs150 (each member)

Veediyawatta village hydro project, Deraniyagala

1. Generation Capacity: 43 kW
2. Total project cost: Rs3 950 000
   a Grants from aid agency (Japan): Rs3 000 000

b Loan from ESD project: Rs800 000
c Community contribution: Rs50 000 (by labour)

3. Number of households – 125, School – 1, Temple – 1
4. Rural economic activities: None
5. Monthly instalment: Rs500 (each member)

Table 2: Use of electrical appliances by households in Kithulritiella village

HH No.	Bulbs	CFL Bulbs	Iron	TV (colour)	TV (b&w)	Cassette/ radio	Heater	Blender	Other
01	6 × 40W	1 × 9W	1	–	12"	–	–	–	–
02	6 × 40W	2 × 5W	1	–	14"	1	1	–	–
03	–	–	–	–	–	–	–	–	–
04	3 × 40W 2 × 40W floracent	4 × 5W	1	20"	–	1	1	1	–
05	4 × 40W 1 × 25W		–	–	14"	1	–	–	–
06	6 × 25W	2 × 12W	1	20"	–	1	1	1	–
07	5 × 40W 1 × 25W	–	1	12"	–	1	–	1	Battery charger (5) + VCD
08	2 × 100W 2 × 40W	–	1	21"	–	1	1	–	–
09	1 × 100W 2 × 40W	2 × 9W	1	–	14"	1	1	1	Amplifier + VCD + Soldering iron
10	7 × 40W	–	1	20"	–	1	1	–	Battery charger + coconut scraper
11	5 × 40W	3 × 9W	1	21"	–	–	1	–	–
12	3 × 60W 2 × 40W	1 × 9W	1	–	14"	1	–	–	–
13	5 × 40W	1 × 7W	1	20"	–	1	1	1	1
14	4 × 40W	–	1	Burnt	–	–	1	–	–
15	3 × 40W 1 × 40W floracent	2 × 9W	1	14"	–	1	1	1	1

Appendix 6: The Cancún negotiations

There are many interpretations of why the Cancún negotiations failed to reach an agreement.[83]

The Director General of the WTO, Supachai Panitchpakdi, was understandably unwilling to blame particular sides, but spoke of 'the Cancún lesson that when participants take too long to unveil their true positions, compromise becomes even more difficult to achieve'. He also spoke of the 'need to work closely with groups of countries and address their concerns earlier to prevent the unnecessary hardening of positions that complicates the decision-making process at ministerial conferences'.[84]

The European Commission's Trade Commissioner, Pascal Lamy tried to be diplomatic but was critical of the G20 countries and the WTO organization. Agence France Presse reported:[85] 'On the farm subsidy issue that appeared to be the main stumbling block in World Trade Organization, Lamy said the EU "showed its readiness to move across the board" and had arrived at a "joint policy framework" with the United States. "Not at all a defensive alliance, in fact almost the reverse..." "But a group of nations led by Brazil and India wanted to pull the text in a different direction. We will never know how far we would have got if the negotiations had continued for another day or so. But I really felt that we were on the verge of a major breakthrough." And 'Lamy was also critical of the current structure of the WTO, which he felt could not handle "the weight and complexity of the negotiating issues with 148 members operating under consensus".'[86]

The American trade representative, Robert Zoellick, was critical of what he saw as the unwillingness of some (developing) countries, which he characterized as 'won't do' to negotiate compared to the 'can do' attitude of the USA, the EU, Canada, and Japan.[87] He suggested that in future, the USA would concentrate on bilateral talks to further open trade.

The Indian trade minister, Arun Jaitley, was more positive about the outcome of Cancún. He stated:[88]

'Cancún failed to produce a document, but it succeeded in focusing on trade distortions in agriculture. Cancún also indicated that the WTO, instead of being driven by only a few, will have to be more participatory. Both of these developments are very positive for the WTO.' And 'The group of 22 (G-22) is a combination of offensive and defensive interests on the part of its members. It has enabled developing countries to raise their voices and be heard. Alone, our capacity to be heard was not so large.'

The Brazilian government saw the talks as a victory for developing countries. President Luiz Inacio da Silva described it 'as a historic moment, in which poor countries managed to block commercial victories of rich countries'.[89] The Brazilian agriculture minister, Roberto Rodrigues, blamed developed countries:[90] 'The negotiator's efforts were ruined by the intransigent stance of the developed countries'.

In Sri Lanka, President Chandrika Bandaranaike Kumaranatunga was heavily critical of developed countries for the Cancún failure. Speaking at the 12th East Asia Economic Summit, she said: 'We do not comprehend how rich nations demand of us to abandon to the whims of the global markets vulnerable sectors of our economy such as the farmers and small industrialists when they practise extensive protectionist policies in their countries'. She also expressed hope that growing trade links with India and the rest of the subcontinent would position Sri Lanka as a services hub for the region, noting that these links had already boosted investment in a wide range of the country's industries, from telecommunications to tourism.[91]

Notes and references

EXECUTIVE SUMMARY

1. http://www.imf.org/external/np/loi/2001/lka/01/index.htm
2. http://www.imf.org/external/np/sec/pr/2003/ pr0354.htm
3. http://www.southcentre.org/info/southbulletin/ bulletin80/bulletin80-03.htm
4. World Bank (2004) 'Credible Regulation Vital For Infrastructure Reform To Reduce Poverty, Says World Bank', press release, June 14, 2004, World Bank, Washington. (see http://web.worldbank.org/WBSITE/EXTERNAL/NEWS/0,,contentMDK:20212113~menuPK:34464~pagePK:64003015~piPK:64003012~theSitePK:4607,00.html)

PROBLEMS WITH THE CONVENTIONAL ELECTRICITY SECTOR IN SRI LANKA

5. 'Sri Lanka Rural Electrification Policy', Ministry of Power and Energy, Colombo (November 2002).
6. In January 2004, US$1 = Rs97.20500 (Sri Lankan Rupee)
7. 'Proposed Power Sector Policy Guidelines', Ministry of Power and Energy, Colombo (2002), p3.
8. *Recent Economic Development: Highlights of 2002 and prospects for 2003*, Central Bank of Sri Lanka (2002).
9. Siyambalapitiya, T., interviewed by Citizens' Trust (2003).
10. The Committee's exercise of powers relating to the Petroleum industry was facilitated by the Petroleum Products (Special Provisions) Act No.33 of 2002.
11. 'An Analysis of the Power Crisis' by Karin Fernando, pub. Citizens' Trust 2003.

RURAL ELECTRIFICATION AND NON-TRADITIONAL ENERGY RESOURCES IN SRI LANKA

12. 'Sri Lanka Rural Electrification Policy', Ministry of Power and Energy, November (2002).
13. ibid
14. ibid, p.4.
15. 'The World Bank Energy Services Delivery Project (IDA-29380) Implementation Completion Report', The World Bank, Washington DC (June 5, 2003).

REFORMS TO THE SRI LANKAN ELECTRICITY INDUSTRY

16. Kessides, I.N., 2004. Reforming Infrastructure Privatization, Regulation, and Competition. A co-publication of the World Bank and Oxford University Press.
17. Ibid.
18. Electricity market reform failures: UK, Norway, Alberta and California Energy Policy, Volume 31, Issue 11, September 2003, Pages 1103–1115 Chi-Keung Woo, Debra Lloyd and Asher Tishler and S Thomas 'The Ontario Government's proposals on electricity restructuring: Comments by Public Service International Research Unit' PSIRU, Greenwich, http://www.psiru.org/reports/2004-08-E-Ontario.pdf
19. There is already considerable private sector involvement in power generation (see section on solar power in previous chapter).
20. Joint letter dated 20.06.2003 from CT/ITDG/EF to reform authorities following Workshop of 8 May, 2003.
21. A Petroleum Reform Act is being drafted, and the CPC monopoly over public petrol sale centres has already been ended.
22. PUCSL Act Section 1.
23. Professor Rohan Samarajiva, at 8 May, 2003 Workshop .
24. PUCSL Act Sections 17 and 30.
25. Usually the prime minister.
26. PUCSL Act Section 14(2).
27. Workshop, 8 May 2003.
28. 'World Development Report 2000/2001: Attacking poverty', Oxford University Press, Oxford and New York (available at www.worldbank.org/poverty/wdrpoverty/report/toc.pdf).
29. Y Albouy & R Bousba (1998) 'The

Impact of IPPs in Developing Countries; Out of the Crisis and into the Future' Viewpoint. Note no 162, World Bank, Washington.
30. R David Gray & J Schuster (1998)' The East Asian Financial Crisis- Fallout for Private Power Projects' Viewpoint. Note no 146, World bank, Washington.

THE WORLD TRADE ORGANIZATION AND THE GATS NEGOTIATIONS

31. The original 21 countries were: Argentina, Bolivia, Brazil, China, Chile, Colombia, Costa Rica, Cuba, Ecuador, Egypt, Guatemala, India, Mexico, Pakistan, Paraguay, Peru, Philippines, South Africa, Thailand, Venezuela and El Salvador. El Salvador left during the Cancún Summit but Nigeria joined.
32. www.wto.org/english/thewto_e/whatis_e/ tif_e/fact2_e.htm
33. See www.wto.org
34. Previous Ministerial Conferences were held in Singapore (December 1996), Geneva (May, 1998), Seattle (December 1999), and Doha (November, 2001).
35. Cho, A. and Dubash, N., *Will investment rules shrink policy space for sustainable development? Evidence from the electricity sector*, World Resources Institute Working Paper, Washington DC (2003).
36. www.wto.org/english/tratop_e/serv_e/energy_e/energy_e.htm
37. For a list of sectors, see www.wto.org/english/tratop_e/serv_e/mtn_gns_w_120_e.doc
38. www.wto.org/english/tratop_e/serv_e/requests_offers_approach_e.doc
39. http://tsdb.wto.org/wto/Public.nsf/FSetReportPredifinedAffich?OpenFrameSet&Frame=F_PredefinedReport&Src=_c5trn8rpfa1qm4r39ccn6ssr65ssmcp9l6kp3cdhhcgoj2p9g74om6c9i6kr3gohg60o3co9n69j3abpoccp3ichm70q6co9o6ksj4dhlccoj4d9m74r68c1g6cq3gohl6cvkap39eh26uorldlimst00_
40. For an account of the Special Sessions, see www.wto.org/english/tratop_e/serv_e/s_negs_e.htm
41. The Documents referred to in the following section are posted at www.wto.org/english/tratop_e/serv_e/s_propnewnegs_e.htm
42. For a discussion of the US negotiating position, see Griffin Cohen, M., 'From public good to private exploitation: GATS and the restructuring of Canadian electrical utilities', *Canadian–American Public Policy*, Occasional Paper 48, Bangor, USA (2001).
43. See: www.polarisinstitute.org/gats/main.html and http://www.gatswatch.org/requests-offers.html#outgoing
44. Joy, C. and Hardstaff, P. 'Whose development agenda? An analysis of the European Union's GATS requests of developing countries' World Development Movement, London (2003).
45. The other sectors are: professional, business, telecommunications, postal and courier services, construction, distribution, environmental, financial, tourism, news agency services, and transport.
46. For the full text of the request to Sri Lanka, see www.gatswatch.org/docs/offreq/EUrequests/SriLanka.pdf
47. www.gatswatch.org/requests-offers.html
48. Australia, Bahrain, Canada, Iceland, Japan, Liechtenstein, New Zealand, Norway, Panama, Paraguay, South Korea, Switzerland, Taiwan, the United States and Uruguay
49. www.dfat.gov.au/trade/negotiations/gats_schedule_initial_offer_0303.pdf
50. strategis.ic.gc.ca/epic/internet/instp-pcs.nsf/vwGeneratedInterE/sk00079e.html
51. www.gatswatch.org/docs/offreq/offers/NewZealand.pdf
52. www.ustr.gov/sectors/services/2003-03-31-consolidated_offer.pdf
53. www.odin.dep.no/archive/udvedlegg/01/05/ wto02057.pdf
54. www.gatswatch.org/docs/offreq/EUoffer/EU-initialoffer.pdf
55. For detailed criticisms of GATS, see the World Development Movement

website (www.wdm.org.uk/campaign/GATS.htm#GATSreports), the GATSwatch website set up by Corporate Europe Observatory and Transnational Institute (www.gatswatch.org/), and the Seattle to Brussels Network (www.wdm.org.uk/cambriefs/gatsdemo.pdf).

56. The arguments presented are summarized from WTO 'GATS – Fact and fiction', www.wto.org/english/tratop_e/serv_e/gatsfacts1004_e.pdf
57. The arguments presented are summarized from World Development Movement 'Stop the GATSastrophe', www.wdm.org.uk/cambriefs/wto/stopgats.pdf
58. Article XIX, General Agreement on Trade in Services http://www.wto.org/english/docs_e/legal_e/26-gats.pdf]. In its introduction to the GATS, the WTO states: [http://www.wto.org/english/thewto_e/whatis_e/tif_e/agrm6_e.htm.
59. www.wto.org/english/thewto_e/whatis_e/ tif_e/agrm5_e.htm
60. www.wto.org/english/tratop_e/serv_e/s_propnewnegs_e.htm#energy
61. www.wto.org/english/tratop_e/serv_e/s_propnewnegs_e.htm#energy
62. www.wto.org/english/thewto_e/minist_e/ min03_e/min03_14 sept_e.htm
63. This meeting was required by the Ministerial Statement at the end of the Cancún conference which stated (Item 4) 'We therefore instruct our officials to continue working on outstanding issues with a renewed sense of urgency and purpose and taking fully into account all the views we have expressed in this Conference. We ask the Chairman of the General Council, working in close co-operation with the Director-General, to co-ordinate this work and to convene a meeting of the General Council at Senior Officials level no later than 15 December 2003 to take the action necessary at that stage to enable us to move towards a successful and timely conclusion of the negotiations. We shall continue to exercise close personal supervision of this process.'

APPENDIX 1: RETREAT OF MULTINATIONAL ELECTRICITY COMPANIES

64. Bayliss, K. and Hall, D. 'Enron: A corporate contribution to global inequality', PSIRU, London (2001) (www.psiru.org/reports/2001-12-enronpro.doc).
65. Most of Enron's non-US assets (in 14 countries) are being transferred into a new holding company expected to be known as Prisma Energy International.

APPENDIX 3: PERCEPTIONS OF THE SRI LANKAN ELECTRICITY INDUSTRY

66. By Citizens' Trust for 'An Analysis of the Power Crisis' by Karin Fernando, pub. Citizens' Trust 2003.
67. Citizens' Trust, 'An Analysis of the Power Crisis' by Karin Fernando, pub. Citizens' Trust 2003.
68. Most of the households using batteries and kerosene were not supplied by the grid. As the majority of the surveys were carried out in urban areas, with few rural areas, it was not possible to capture the use of alternatives accurately.
69. Joint letter dated 20.06.2003 from CT / ITDG / EF to reform authorities following Workshop of 8 May 2003.
70. Electricity Reform Act No.28 of 2002, Section 12(4).
71. ibid, Section 9(1).
72. Thirteenth Amendment to the Constitution.
73. The Head of the Public Interest Program Unit, Professor Rohan Samarajiva, suggested that such co-ordination could be achieved through MOUs entered into between the PUC and the provincial authorities. Workshop conducted by Citizens' Trust (CT), Intermediate Technology Development Group (ITDG) and Energy Forum (EF), 8 May 2003.

74. A clear indication of this development was provided when the World Bank in June 2003 called for consultants to do a study of the impact of the proposed reforms on its RERED scheme.
75. Interviews by Citizens' Trust for *Analysis of Sri Lanka's Power Crisis*.

APPENDIX 4: THE CEB GENERATION PLAN: 2002–16

76. The mathematical computer package used is called WASP – Wien Automatic System Planning Package, developed by the International Atomic Energy Authority.
77. The optimization cost is derived using system costs, economical costs and outage costs.
78. Also referred to as 'least cost options'
79. Effectively the real (net of inflation) cost of capital.
80. Interviews with Citizens' Trust for *Analysis of Sri Lanka's Power Crisis*.
81. Nor has the Plan been revised to accommodate changed government thinking on the location of the proposed coal power plant originally earmarked for the West Coast.
82. Citizens' Trust, *An Analysis of Sri Lanka's Power Crisis*.

APPENDIX 6: THE CANCÚN NEGOTIATIONS

83. For a developing country account of the negotiations see *South Bulletin*, September 30, 2003. www.southcentre.org/info/southbulletin/bulletin64-65/toc.htm
84. www.wto.org/english/news_e/news03_e/news_sp_18sep03_e.htm
85. Agence France Presse, September 23, 2003, 'WTO Doha Round on "life-support" says EU Trade Commissioner Lamy'.
86. For a fuller account of Lamy's views, see http://europa.eu.int/comm/commissioners/lamy/speeches_articles/spla195_en.htm
87. Zoellick, Robert, 'America will not wait for the won't-do countries', *Financial Times*, (September 22, 2003).
88. *Business Week*, September 23, 2003 'Where the Cancún Talks "Succeeded"; Indian Commerce & Industry Minister Arun Jaitley says after the breakdown in Mexico, the WTO "will have to be more participatory".
89. Gazeta Mercantil, September 16, 2003.
90. *World News Connection*, September 16, 2003, 'Brazil: Agriculture minister believes Seattle, Cancún failure put WTO at stake'.
91. *Malaysian Economic News*, October 14, 2003, 'Sri Lankan President calls for WTO overhaul'.